Throne
Understanding Thermoforming

Hanser **Understanding** Books

A Series of Mini-Tutorials

Series Editor: E.H. Immergut

Understanding Injection Molding Technology (Rees)
Understanding Polymer Morphology (Woodward)
Understanding Product Design for Injection Molding (Rees)
Understanding Design of Experiments (Del Vecchio)
Understanding Plastics Packaging Technology (Selke)
Understanding Compounding (Wildi/Maier)
Understanding Extrusion (Rauwendaal)
Understanding Thermoforming (Throne)

James L. Throne

Understanding Thermoforming

Hanser Publishers, Munich

Hanser Gardner Publications, Inc., Cincinnati

The Author:
Dr. James L Throne, Sherwood Technology, Inc., 158 Brookside Blvd, Hinckley, OH 44233-9676, USA

Distributed in the USA and in Canada by
Hanser Gardner Publications, Inc.
6915 Valley Avenue, Cincinnati, Ohio 45244-3029, USA
Fax: (513) 527-8950
Phone: (513) 527-8977 or 1-800-950-8977
Internet: http://www.hansergardner.com

Distributed in all other countries by
Carl Hanser Verlag
Postfach 86 04 20, 81631 München, Germany
Fax: +49 (89) 98 12 64

Die Deutsche Bibliothek – CIP-Einheitsaufnahme

Throne James L.:
Understanding thermoforming/ James L. Throne. – Munich; Vienna;
New York: Hanser; Cincinnati: Hanser Gardner, 1999
ISBN 3-446-21153-5

Typeset in the U.K. by Marksbury Multimedia Ltd, Bath
Printed and bound in Germany by Druckerei Wagner, Nördlingen

Forword

This is a primer on thermoforming. The thermoforming industry is growing much more rapidly than the plastics processing industry as a whole. This growth has been stimulated by the commercialization of:

- Pressure forming
- Twin-sheet forming
- Advanced heater design
- Polymers designed for the thermoforming process
- New mold materials and techniques
- Advanced robotics
- Reliable temperature monitoring and control
- Computer-aided design techniques

This growth has also attracted the interest of plastic products manufacturers who consider the thermoforming process as competition for their current plastics processes, such as injection molding and blow molding; who desire to add thermoforming to the processes they currently run; or who need to know more about the thermoformed parts they are currently purchasing.

This primer is a compilation of materials presented at many in-plant short courses and the basic thermoforming courses sponsored by University of Wisconsin-Milwaukee. It is designed for people who may not have extensive formal technical education such as plant supervisors and foremen, machine operators, quality control people, purchasing agents, and technical secretaries. The primer includes a glossary of thermoforming terms, a recommended reading list for more information, and an extensive subject index. There are no equations.

James L. Throne

Contents

1 Introduction to Thermoforming

Annual plastics consumption in the US is approaching 100 million pounds or 45 Mkg. Plastics are converted from resin into useful products in many ways. Injection molding and extrusion are the primary conversion technologies. Thermoforming is a minor conversion technology, as are blow molding, rotational molding, and many other processes. This chapter outlines the history of thermoforming technology, general markets, and processing methods technologies that compete with thermoforming.

1.1 Brief History

In thermoforming, plastic articles are shaped from softened sheet. The process involves heating the plastic sheet to a temperature range where it is soft, and then stretching the softened sheet against a cool mold surface. When the sheet has cooled to the point where it retains the shape of the mold, the sheet is removed from the mold and excess plastic trimmed from the part.

Thermoforming is a generic term for a group of processes including vacuum forming, drape forming, billow or free bubble forming, mechanical bending, matched-mold forming, and the newer processes of pressure forming and twin-sheet forming.

Thermoforming is one of the oldest methods of forming useful articles. In the 1870s in the US, John Wesley Hyatt, considered the father of modern plastics processing, and his colleague, Charles Burroughs, rolled thin, skived sheets of celluloid or cellulose nitrate into tubes, inserted them into steel pipe that contained the desired shape, and heated the polymer with steam under pressure. The heat softened the celluloid and forced it against the pipe shapes. The pipes were then cooled in water, thereby rigidifying the celluloid in its final shape, such as a baby rattle or a small, shaped bottle. Table 1.1 provides highlights of other advances in thermoforming.

Table 1.1 Brief History of Thermoforming

Time	Place	Thermoforming Activity
Prehistory -	Egypt	Tortoise sheet, keratin, is heated in hot oil and shaped to produce food containers
Prehistory -	Americas	Tree bark, natural cellulosics, heated in hot water and shaped into bowls and canoes
1870s	US	Hydraulic planer developed for cutting thin sheets of celluloid, Charles Burroughs Co., NJ
1910	England	Sharps piano keys drape-formed over captive wooden cores
1930	US	Bottle formed from two thermoformed halves by Fernplas Corp
1930s	US	Relief maps thermoformed for US Coast & Geodetic Survey
1933	Europe	Formed rigid PVC used as liner in Phillips refrigerator
1935	US	Cellulose acetate ping-pong balls twin-sheet formed by E.I. DuPont de Nemours Co., Inc., Leominster, MA
1938	US	Blister pack of cellulose nitrate.
1938	Germany	Automatic thin-sheet roll-fed thermoformer developed by Klaus B. Strauch Co.
1938	US	Cigarette tips, ice-cube trays automatically thermoformed
1942	UK, US	Cast PMMA acrylic thermoformed for fighter/bomber windows, gun closures, windscreens
1948	UK	Cast PMMA acrylic bathtubs thermoformed by Troman Brothers
1954	US	Skin-packaged products shown at Hardware Manufacturers Association, Chicago
1970	US	Thermoformed ABS "concept car" automobile body by Borg-Warner, Inc.

1.2 General Markets

Thermoformed products are typically categorized as permanent or industrial products and disposable products. Often, disposable thermoformed products are used in packaging.

Typical industrial products include equipment cabinets for medical and electronic equipment, tote bins, single and double deck pallets, transport trays, automotive inner-liners, headliners, shelves, instrument panel skins, aircraft cabin wall panels, overhead compartment doors, snowmobile and motorcycle shrouds, farings and windshields, marine seating, lockers and windshields, golf carts, tractor, and RV shrouds, skylights, shutters, bath and tub surrounds, lavys, storage modules, exterior signs, swimming and wading pools, landscaping pond shells, luggage, gun and golf club cases, boat hulls, animal carriers, and seating of all types.

Typical disposable products include blister packs, point-of-purchase containers, bubble packs, slip sleeve containers, audio/video cassette holders, hand and power tool cases, cosmetic cases, meat and poultry containers, unit serving containers, convertible-oven food serving trays, wide-mouth jars, vending machine hot and cold drink cups, egg cartons, produce and wine bottle separators, medicinal unit dose portion containers, and form-fill-and-seal containers for foodstuffs, hardware supplies, medicine, and medicinal supplies.

Major thermoforming growth areas include multilayer containers, retortable containers, modified atmosphere packages, transportation and transport refurbishment, medical device equipment, and dunnage.

According to recent market surveys, the world market for thermoforming is growing at around 7 to 10% per year, on a weight basis. A 1995 survey of North American industrial thermoforming estimated that market to be about 1500 million pounds (700 million kg) in 1994, with a projected growth to 2200 million pounds (1000 million kg) by 2000. The thin-gauge market is estimated to be twice the size of the industrial thermoforming market. As a result, it is estimated that the total North American thermoforming market will be about 6600 million pounds [3000 million kg] by the year 2000, with an estimated value of about US $15 billion.

The European thermoforming market is approximately 60% of the North American market; the Asian market about 40%; the South American market about 20%; and the rest of the world 20% of the size of the North American market. From this information, the estimated size of the world-wide thermoforming market, including industrial and thin-gauge products, in the year 2000 is about 16,000 million pounds [7200 million kg], with a value of about US $35 billion.

1.3 Terminology

As noted earlier, there are discrete steps in thermoforming. The sheet is first clamped, then heated. Once the sheet is soft enough for forming, it is shaped by bringing it in contact with a mold. When it is cool enough to retain the shape of the mold, the sheet is removed and the product is trimmed from the excess material around it.

Table 1.2 Characteristics of Thin-Gauge and Heavy-Gauge Thermoforming

Characteristic	Thin-Gauge	Heavy-Gauge
Initial sheet thickness	<0.060 in or 1.5 mm	>0.120 in or 3 mm
Dominant products	Packaging, disposables	Cabinetry, industrial
Sheet handling	Roll-fed	Palletized cut sheet
Typical machine type	In-line extruder-former	Shuttle or rotary press
Machine control aspects	Automated	Automated to manual
Controlling aspect, heating	Heater output, W/in^2 or W/m^2	Conduction into sheet
Pattern heating	Difficult, usually not done	Common
Part size tendency	Small	Medium to very large
Number of mold cavities	Many	One or two, usually
Mechanical assist	Plug	Plug, billow, vacuum box
Mold type	Female, usually	Male, female, androgynous
Mold materials	Aluminum, machined	Wood, plaster, syntactic foam, white metal, cast aluminum
Mold cooling	Actively controlled	Active to none for prototype
Free surface cooling	Ambient, usually	Forced air, fogging
Trimming aspects	Punch and die, rim rolling	Multi-axis routing
Non-product trim level	About 50%	About 25% to 30%
Wall thickness tolerance, normal	20%	20%
Wall thickness tolerance, tight	10%	10%
Pressure forming application	Deep draw, formed rim	Textured surfaces, deep draw

Typically, thermoforming processes are loosely sub-divided according to the thickness or gauge of the sheet used. When the sheet is less than 0.060-inch (60 mils or 1.5 mm) thick, the process is called thin-gauge thermoforming. When the sheet is less than about 0.010-inch (10 mils, 0.25 mm or 250 μm) in thickness, it is often called film or foil. Heavy-gauge thermoforming is used when the sheet is greater than about 0.120-inch (120 mils or 3 mm) in thickness. When the sheet is greater than about 0.500-inch (500 mils or 13 mm) in thickness, it is often called a plate.

Another way of dividing thermoforming is by the way the sheet is presented to the thermoforming press. If the sheet is thin enough, it is usually in rolls of up to 40 to 60 inches (1 to 1.5 m) in diameter, weighing as much as 5000 pounds (2300 kg), and containing as much as 10 000 feet (3000 m) of sheet. This sheet is fed continuously into roll-fed thermoforming machines. If the sheet is too thick to be rolled, it is usually guillotined into discrete pieces, which are then stacked on pallets. These sheets are then fed, either manually or automatically, into cut-sheet thermoforming machines.

There is a "gray area" between thin-gauge and heavy-gauge thermoforming. Usually the key to processing sheet in this thickness range between the two is whether the sheet is too stiff to roll and must therefore be cut and palletized to ship it. Polystyrene and polyolefin foam sheet thicknesses are usually greater than 0.120-inch (3 mm), but these foams are soft enough to roll and are therefore treated as thin-gauge sheet stock. Table 1.2 summarizes some characteristics of thin-gauge and heavy-gauge thermoforming.

1.4 The Competition to Thermoforming

Three or four major plastics processes compete with thermoforming: blow molding, injection molding, rotational molding and, in certain instances, glass-fiber reinforced (GR) thermoset molding. For thin-gauge, disposable, rigid containers, the primary competition is blow molding for hollow containers and injection molding for wide-mouth containers. For the production of heavy-gauge, non-cosmetic parts, the primary competition is blow molding and rotational molding, with injection molding as the primary competition in cosmetic parts. GR molding is competitive only when the formed part must exhibit high mechanical performance. General comparisons of four thermoplastic processes are given in Table 1.3.

Thermoforming is a low-temperature, low-pressure process. Furthermore, it needs only a single-surface mold. As a result, relatively inexpensive mold materials are used when the production runs are short. Additionally, molds are fabricated in relatively short times, which allows customers to "proof" their designs, field test products, and approve prototypes for production in a relatively short time. Because thermoforming begins with sheet, there is no need for plastic to "flow" from one point to another, as it must in most converting processes. As a result, with thermoforming, parts can be produced with a very high surface-to-thickness ratio. As an example, 0.6 mil (15 μm) thick, oriented polyethylene terephthalate [OPET] can be thermoformed into very precise 2-inch (50 mm) diameter microphone

Table 1.3 Comparison of Various Thermoplastic Processes

Characteristic	Thermoforming	Injection Molding	Blow Molding	Rotational Molding
Polymer form	Sheet	Pellets	Pellets	Powder
Variety of polymer	Good to excellent	Excellent	Good	Fair to limited
Raw material cost	Polymer + extrusion	Standard	Standard	Polymer + grinding
Variety of mold materials	Very many	Very limited	Limited	Many
Mold cost	Moderate to low	Highest	High	Moderate to low
Production mold material	Aluminum	Steel	Aluminum	Aluminum, steel
Thermal cycling of mold	Gentle	Moderate	Moderate	Severe
Part wall uniformity	Fair to poor	Excellent	Poor to fair	Good to excellent
Major design problems	3D corner, wall thickness uniformity	Gating, weld line	Pinch-off, wall uniformity	Porosity
Part failure mode	Thin corners, microcrack	Weld line walls, poor pinch-off	Thin side	Poor tensile strength
Operating pressure, atm	−1 to 5	100 to 1000	5 to 25	0 to 1
Operating temperature, °C	to 200	150 to 300	100 to 250	200 to 350
Filling methods	Manual to automatic	Automatic	Automatic	Manual
Part removal methods	Manual to semi-automatic	Automatic	Automatic	Manual
Flash, trim	Highest	Low to very low	Moderate to high	Moderate to low
Inserts	Possible	Feasible	Feasible	Feasible
Orientation in part	Highest	Moderate to high	High to moderate	Unoriented
Stress retention	Highest	High	High to very high	None to little
Part surface finish	Good to excellent	Excellent	Very good	Good
Surface texture	Good to very good	Excellent	Very good	Good to fair

diaphragms, with an area-to-thickness ratio of 130 000. No other process can produce similar results.

Thermoforming has several disadvantages over other processes, however. Thermoforming begins with extruded plastic sheet. Blow molding and injection molding begin with plastic pellets. The extra initial step adds cost, typically more than US $0.15 per pound (US $0.33 per kilogram), to the final product. Then, to form a part, the sheet must be held during forming. The plastic in this region, called trim, web, or skeleton, must be trimmed away from the part. Although this plastic is usually reground and re-extruded, it can be thermally damaged by the first extrusion, thermoforming, and grinding processes. The trimmed material is also generally considered to have lower physical properties and economic value than virgin polymer. In addition, this regrind must be re-extruded, again at an expense of more than US $0.15 per pound (US $0.33 per kilogram). Issues related to the use of regrind are discussed in more detail in Chapter 10.

Finally, thermoforming is typically a "one-sided" process. In other words, unlike injection molding where the plastic is squeezed between two mold halves, in thermoforming, the hot, formable plastic sheet is forced against a single-sided mold. The other surface of the sheet usually is allowed to cool without touching a shaping surface. While this means the molds used in thermoforming are relatively inexpensive, the progressive laying of the sheet against the mold surface yields a part that does not have uniform wall thickness. As a result, wall thickness tolerance is typically 20%, or similar to that for blow molded and rotational molded products. These two processes are also "one-sided." In contrast, wall thickness tolerance for injection molded products is typically less than 3% and frequently less than about 1%.

Wall thickness variation can be controlled to some extent by manipulating the formable sheet prior to forcing it on the mold, but injection molded wall tolerance cannot be achieved in thermoformed products. As a result, thermoformers must design to the minimum allowable critical thickness, meaning that many portions of the formed part contain more plastic than required. This means that, in general, thermoformed parts tend to be designed conservatively and can be more costly than expected. Part design is considered in Chapter 9.

2 Polymers and Plastics

Thermoforming uses plastic sheet, which is heated, stretched, cooled, and mechanically cut. For the most part, the plastic sheet is manipulated as a rubbery solid or elastic liquid. As a result, the solid or elastic liquid properties of polymers are more important in thermoforming than their viscous properties. In this chapter, important polymer characteristics are examined. Then individual polymers are discussed. Filled, reinforced, and foamed polymer characteristics are presented, as well.

2.1 Polymer Characterization

Polymers are organic molecules consisting of long, repeated chains of simple molecules. As an example, polyethylene, the polymer most used in the world, is made by reacting ethylene under high temperature and pressure, and in the presence of a catalyst. Ethylene is a gas with a chemical composition of $H_2C=CH_2$, where C is carbon, H is hydrogen and the symbol "=" indicates a double bond or a reactive link between the carbons. Ethylene is called a monomer, and has a melting temperature of $-169\,°C$ and a boiling temperature of $-104\,°C$.

The chemical structure of polyethylene is sometimes written as $H_2C\text{-}(CH_2\text{-}CH_2)_x\text{-}CH_2$, where "x" represents the number of ethylenic segments or mers in the polymer. If the value of x is relatively small, on the order of 100, the polymer is more like a high-temperature wax. If the value of x is relatively high, on the order of 5000, its melting temperature is about 130 °C and it degrades before it boils. The polymer is generally processed by one of the standard plastics processing techniques, such as injection molding, blow molding, rotational molding, extrusion, and thermoforming. If the value of x is very high, on the order of 300,000, the polymer may be intractable in normal processing machines and techniques such as compression molding and sinter-fusion of powder may be required.

Although practitioners generally consider the words plastics and polymers interchangeable, the term plastics refers to the product delivered as resin pellets or sheet. Nearly all plastics contain polymers, the pure long-chain hydrocarbons, but they also may contain a wide variety of additives such as thermal stabilizers, antioxidants, color correcting dyes, and internal and external processing aids, as well as product-specific additives such as fire retardants, colorants, UV stabilizers, and fillers.

There are two general categories of polymers. When the polymer can be heated and shaped many times without substantial change to its characteristics, it is a thermoplastic. When the polymer cannot be reshaped after being heated and shaped the first time, it is a thermoset. Thermoforming is primarily concerned with thermoplastics.

Thermoformers use two general types of thermoplastic polymers. When a polymer is heated from a very low temperature, it undergoes a transition from its glassy state to a rubbery state. Although this transition occurs over several degrees of temperature, usually only one temperature value is reported as the glass transition temperature. Polymers that only have a glass transition temperature are called amorphous polymers. Polystyrene, ABS, PVC, and polycarbonate are examples of amorphous polymers. Some 80% of all polymers thermoformed are amorphous polymers. About 80% of all amorphous polymers are styrenic, such as polystyrene, impact polystyrene, ABS, and similar materials.

Certain polymers exhibit a second transition from the rubbery state to a molten or melt state. Again, this transition occurs over several degrees of temperature, and again usually only one temperature value is reported as the melt temperature. Polymers that have both glass transition and melt temperatures are called crystalline polymers. Polyethylene and polypropylene are examples of crystalline polymers. Table 2.1 gives transition temperatures for typical thermoformable polymers.

If only one polymer specie is used in a given plastic recipe, the polymer is called a *homopolymer*. Examples of homopolymers include lowdensity polyethylene (LDPE), made entirely of reacted ethylene monomer molecules; general purpose polystyrene (GPPS), which is made entirely of reacted styrene monomer molecules, and is sometimes called *crystal polystyrene* because parts made of the unpigmented water-white polymer have the appearance of fine crystal; and polyethylene terephthalate (PET). Although PET is made by reacting ethylene glycol with terephthalic acid, the resulting polymer has only one type of repeat unit or *mer,* ethylene terephthalate, and therefore PET is a homopolymer. Polycarbonate (PC) is another homopolymer that is made by reacting two types of organic

Table 2.1 Transition Temperatures of Some Thermoformable Polymers

Polymer	Glass Transition Temperature		Melting Temperature		Heat Distortion Temperature 66 lb/in² or 0.46 N/mm²	
	[°F]	[°C]	[°F]	[°C]	[°F]	[°C]
Polystyrene	200	94	–	–	155–204	68–96
PMMA	212	100	–	–	165–235	74–113
PMMA/PVC	221	105	–	–	177	81
ABS	190–248	88–120	–	–	170–235	77–113
Polycarbonate	300	150	–	–	280	138
Rigid PVC	170	77	–	–	135–180	57–82
PETG	180	82	–	–	158	70
LDPE	−13	−25	239	115	104–112	40–44
HDPE	−166	−100	273	134	175–196	79–91
Cellulose acetate	158,212	70,100	445	230	125–200	52–93
Polypropylene	41	5	334	168	225–250	107–121
Co-Polypropylene	−4	−20	302–347	150–175	185–220	85–104
PET	158	70	490	255	120	49

molecules, bisphenol A and phosgene, to produce a polymer with only one type of repeat unit.

If one polymer is reacted with another, the polymer is called a copolymer. Impact polystyrene (HIPS) is an example of polystyrene reacted with a rubber such as butadiene. Many copolymers are used in thermoforming, including polypropylene-polyethylene and PVC-PMMA. If three polymers are reacted together, the polymer is called a terpolymer. The classic terpolymer is ABS, which is a reacted product of acrylonitrile, butadiene, and styrene.

Occasionally two polymers are extrusion- or melt-blended together to make a specific plastic recipe. The classic blended polymer is modified polyphenylene oxide (mPPO), which is a near-equal blend of polystyrene and polyphenylene oxide. mPPO has good impact resistance and fire retardancy properties.

Additives are found in many polymers. Some of these additives are required simply to make a polymer processable. Polyvinyl chloride (PVC) is compounded with many additives to make it processable and useful for many applications. Other additives, called anti-block agents and slip agents, are needed to prevent rolled sheet from sticking together. Additives such as antimony oxide are added when fire retardancy is needed.

By far the majority of thermoplastics that are thermoformed are neat, i.e., they contain no fillers or fibers. Both fillers and fibers reduce polymer formability and as a result, applications are limited to relatively shallow draw, unless substantial pressure is applied during forming. Long fibers and continuous fiber mat require special processing techniques.

2.2 The Thermoforming Window

The thermoforming window is the temperature range over which the polymer is sufficiently subtle or deformable for stretching and shaping into the desired shape. Typically, amorphous polymers have broader thermoforming windows than crystalline polymers. Polystyrene, for example, can be formed from around 260 °F (127 °C), or about 50 °F (30 °C) above its glass transition temperature, to about 360 °F (180 °C) or only a few degrees below the temperature where it is injection moldable. Polypropylene homopolymer, on the other hand, is so fluid above its melting temperature of 330 °F (165 °C) that its thermoforming window may be no more than one degree or so. As a result, it is frequently formed just below its melting temperature. Even then, its thermoforming window may be only two or three degrees.

2.3 Thermal Properties

Plastics are notorious thermal insulators. Since thermoformers need to efficiently heat plastic sheet to a suitable forming temperature, and then cool the formed part to a temperature where the plastic retains the shape of the mold, thermoformers need to know about the thermal properties of polymers. There are three thermal properties that are important:

- enthalpy or heat capacity
- thermal conductivity
- and temperature-dependent density

Table 2.2 gives representative values of these properties for some thermoformable polymers.

2.3.1 Heat Capacity

Heat capacity, sometimes called specific heat, is a measure of the amount of energy required to elevate the polymer's temperature.. The field of study that focuses on the energy uptake of materials is called thermodynamics. In

Table 2.2 Physical Properties of Thermoformable Polymers

Polymer	Density [lb/ft³]	[kg/m³]	Thermal [Btu/] [$\times 10^{-3}$ ft.hr.°F	Conductivity [Btu/lb. kW/ m.°C]	Heat Capacity Coefficient °F or cal/g°C]	Thermal Expansion [$\times 10^{-6}$ °F^{-1}]	[10^{-6} °C^{-1}]
Polystyrene	65.5	1050	0.105	0.18	0.54	40	70
ABS	65.5	1050	0.07	0.12	0.4	50	90
Polycarbonate	74.9	1200	0.121	0.207	0.49	40	70
Rigid PVC	84.2	1350	0.100	0.171	0.365	45	80
LDPE	57.4	920	0.23	0.39	0.95	140	250
HDPE	59.9	960	0.29	0.50	1.05	110	200
PP Homo.	56.2	900	0.11	0.19	0.83	85	150
PET	85.5	1370	0.138	0.236	0.44	40	70
Low-density PS foam	4.0	64	0.016	0.027	0.5	110	200

thermodynamics, one of the fundamental measures of energy uptake is enthalpy. Enthalpy increases with increasing temperature. When a material goes through a characteristic change such as melting, the temperature-dependent enthalpic curve changes dramatically. When a material goes through a characteristic change such as glass-to-rubber transition, the temperature-dependent enthalpic curve changes subtly, if at all.

As expected, it takes far more energy to heat a crystalline polymer from room temperature to above its melt temperature, for example, than to heat an amorphous polymer from room temperature to the same temperature. This is seen in Fig. 2.1. For example, it takes more than twice as much energy to heat polyethylene, a crystalline polymer, to 360 °F (180 °C) than it does to heat polystyrene to the same temperature. Furthermore, because the formed shape must be cooled, twice as much energy must be removed to cool polyethylene to a given temperature than to cool polystyrene to the same temperature.

A single value of specific heat is frequently given for a specific polymer. These values are determined by dividing the enthalpy difference by the temperature difference. Such values are acceptable for amorphous polymers, but should be used with caution concerning a crystalline polymer, because in this case the slope of the temperature-dependent enthalpy curve, and therefore, the specific heat, changes dramatically as the temperature approaches the melt temperature of the polymer.

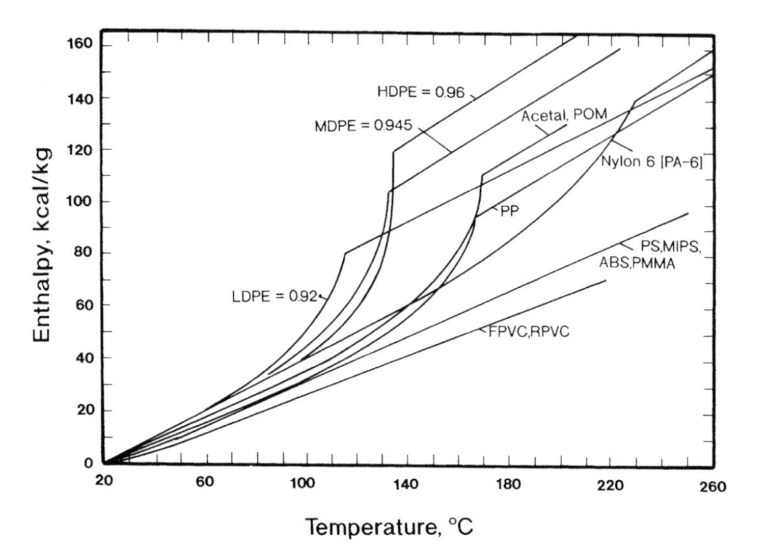

Figure 2.1 Enthalpies of thermoformable polymers.

2.3.2 Thermal Conductivity

Thermal conductivity is the measure of energy transmission through a material. The thermal conductivity values for organic chemicals, including plastics, are in general substantially lower by orders of magnitude than metals, for example. In other words, polymers are thermal insulators. As an example, the thermal conductivity of aluminum, a common metal for thermoforming molds, is nearly one-thousand times greater than the thermal conductivity of polystyrene.

During thermoforming, thermal conductivity is a measure of energy transmission through the polymer sheet. Even though the thermal conductivities of polymers are low, there are differences in these values among polymers. For instance, the thermal conductivity of HDPE is about four times higher than that of polystyrene or ABS. Thermal conductivity and its companion property, thermal diffusivity, discussed below, are quite important when forming very thick sheets, because the rate of energy transfer into the sheet governs, to a large extent, the formability of the sheet. Although thermal conductivity typically decreases slightly with increasing temperature, for most processing purposes the value can be considered constant.

2.3.3 Density

Polymer density decreases and its reciprocal, specific volume, increases with increasing temperature. In the vicinity of the glass transition temperature, the slope of the temperature-dependent specific volume curve changes perceptively. In the vicinity of the melt temperature, the slope changes dramatically. Typically, the density of an amorphous polymer at its forming temperature is about 10 to 15% less than that at room temperature. The density of a crystalline polymer at its forming temperature may be as much as 25% less than that at room temperature. Obviously, as the polymer cools from its forming temperature, its density increases, and volume decreases; as a result, the final part dimensions decrease and the part exhibits shrinkage. This is discussed in Chapter 9.

2.3.4 Thermal Diffusivity

Thermal diffusivity is a combination of polymer properties, and is calculated by dividing the polymer thermal conductivity by its density and specific heat. Thermal diffusivity is the fundamental polymer property in time-dependent heat transfer to materials. Because of the unique bundling of temperature-dependent characteristics of the polymer properties, thermal diffusivity is nearly independent of temperature for nearly all polymers.

2.4 Infrared Energy Absorption

Most commercial thermoforming heaters emit energy in far infrared wavelengths. Figure 2.2 shows the entire electromagnetic radiation band, from the very short radio and microwave wavelengths to the extremely long wavelengths of nuclear and cosmic rays. The visible electromagnetic wavelength spectrum is very narrow, from about 0.4 μm to about 0.7 μm. The near-infrared wavelength spectrum is from about 0.7 μm to about 2.5 μm. The far-infrared wavelength spectrum is from about 2.5 μm to about 100 μm.

Thermoforming is concerned most with the wavelength range from about 2.5 μm to about 15 μm. The chemical make-up of a polymer dictates how much energy a polymer absorbs, and in contrast, how much energy a polymer transmits. Figures 2.3, 2.4, and 2.5 are transmission spectra for polystyrene, polyethylene, and PVC, respectively. There is a significant

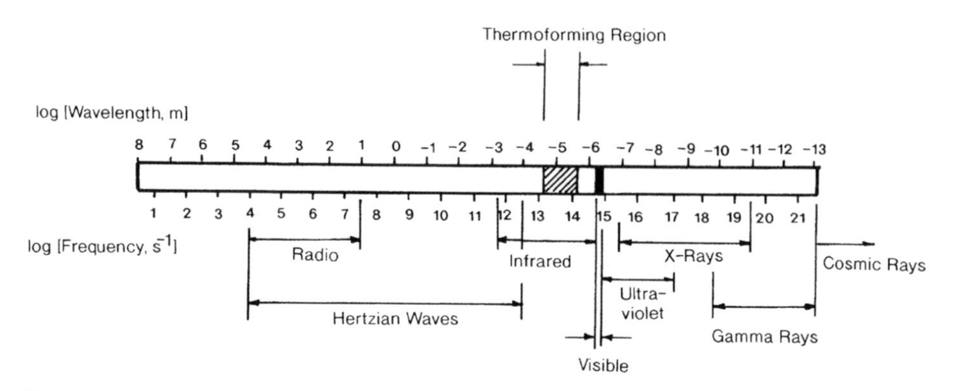

Figure 2.2 Electromagnetic radiation, including visible and infrared regions.

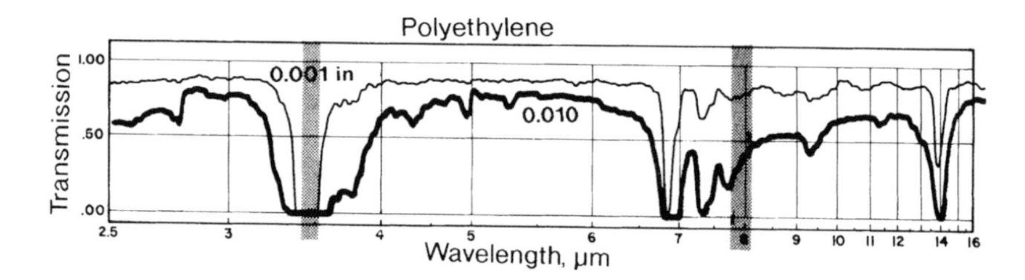

Figure 2.3 Infrared transmission spectrum for polyethylene (PE).

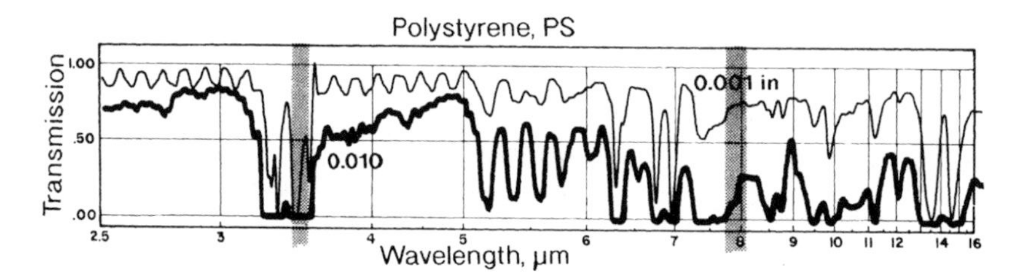

Figure 2.4 Infrared transmission spectrum for polystyrene (PS).

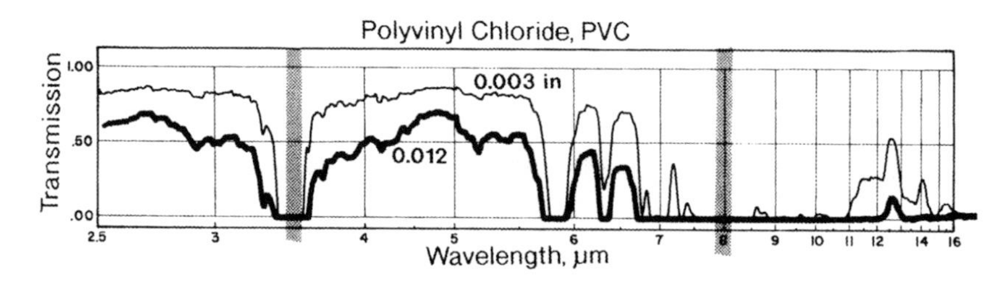

Figure 2.5 Infrared transmission spectrum for polyvinyl chloride (PVC).

difference in the wavelength-dependent transmissions of these polymers. Infrared energy absorption is very important when heating thin sheet and film. As discussed Chapter 4, it is also important when trying to determine energy uptake in thick sheet.

2.5 Thermoformable Polymers

In this section, we examine in detail some of the polymers that are commonly thermoformed.

2.5.1 Polystyrene [PS and Other Styrenics]

Less than 20 years ago, polystyrene (PS) and the family of styrenics such as HIPS, ABS, SAN, ABA, and OPS, dominated the thermoforming industry. For example, in 1983–1984, nearly 80% of all thermoformed products were styrenics. Unmodified PS is characterized by high modulus, low room temperature elongation at break, excellent clarity, superior drawability, and a very broad thermoforming window. Although PS is easy to thermoform, its trim dust is tenacious and improper trimming can cause edge microcracks, which ultimately lead to cracking and part failure.

To improve its impact resistance, PS is either melt-blended or co-reacted with butadiene, a synthetic rubber, to form impact PS (HIPS). While rubber modification allows HIPS to be used in impact-resistant applications such as equipment cabinets, the rubber reduces the modulus and, to some degree, the formability of HIPS. Acrylonitrile, when copolymerized with PS, yields styrene-acrylonitrile (SAN), a very tough, rubber-free polymer with exceptional clarity. When butadiene and acrylonitrile are co-reacted with PS, the result is ABS. ABS is a very versatile polymer used in applications where high impact strength and toughness are required, such as in equipment cabinets, appliances, and truck cab interiors. Since rubber-modified styrenics yellow when overheated, care must be taken to avoid excessive heating during thermoforming. Additionally, ABS sheet, in particular, should be kept dry to prevent moisture bubbles.

Oriented polystyrene (OPS) is used primarily for quality deli containers, such as cake covers. OPS is made by stretching polystyrene sheet in both the machine direction and the cross-machine or transverse direction in a post-extrusion step. The result is a very tough sheet with outstanding clarity. OPS must be heated carefully to minimize loss in orientation. Contact heat thermoformers are usually recommended, particularly for disposable products that can tolerate the substantial cost for the sheet product.

2.5.2 Polyvinyl Chloride [PVC, Vinyl, RPVC, FPVC]

Rigid polyvinyl chloride (RPVC) is compounded with just enough processing aid to allow it to be extruded without substantial thermal damage. RPVC is thermoformed into building products such as window fascia. Flexible PVC (FPVC), on the other hand, requires the incorporation of many processing aids so that it can be formed, yet remain soft and supple. Interior automotive applications include door panels and instrument panel skins. Semi-rigid PVC is frequently calendered into thin-gauge sheet for packaging applications.

Care must be taken to avoid overheating PVC during thermoforming. The first indication of excessive heat is discoloration. Ultimately, PVC degrades to a dark brown color along with the generation of HCl gas, which, although a corrosive acid and an irritant to the mucus membranes, is not toxic. PVC is melt-compatible with polymethyl methacrylate (PMMA), producing a thermoformable polymer with good thermal stability and excellent weatherability, scratch resistance, and fire-retardancy.

2.5.3 Polymethyl methacrylate [PMMA, Acrylic]

Polymethyl methacrylate (PMMA) is the polymer of choice for transparent and translucent sky domes and outdoor signs. It can be drawn better and more consistently than any other polymer.

There are two methods of making PMMA sheet. Cell-cast PMMA sheet is made by pouring reactive resin in a frame, then warming the assembly to allow the resin to react. Extruded PMMA sheet is produced via conventional extrusion of acrylic pellets. Cell-cast PMMA has much greater molecular weight than extruded PMMA, and it is tougher, more scratch-resistant, and more difficult to thermoform. Care must be taken to ensure the sheet is hot enough when forming into three-dimensional corners. Cold-formed corners can be brittle and can exhibit severe stress-cracking. Although not a critical problem, absorbed moisture can cause PMMA to haze during heating in a thermoforming machine.

2.5.4 Cellulosics [Celluloses, CN, CAB, CAP]

In the early part of the 20th century, cellulosics were the dominant formable polymers. There are many types of cellulosics. Cellulose nitrate (CN) was the earliest of this group to be processed through thermoforming and is no longer manufactured. Its first replacement, cellulose acetate (CA), has itself been replaced for the most part by cellulose acetate butyrate (CAB) and cellulose acetate propionate (CAP). Celluloses are essentially natural polymers that have been chemically altered to improve specific properties. Although cellulosics are relatively easy to thermoform, they absorb moisture, that can cause pinholes in the final product unless the sheet is kept very dry. Furthermore, they are relatively expensive and so are finding fewer applications in their original area, disposable packaging.

2.5.5 Polycarbonate [PC]

Polycarbonate (PC) is well-known as a very tough, high temperature, transparent plastic. It, like PMMA, finds use in signage and skylights. Unlike PMMA, however, it yellows under long-term exposure to UV radiation. Also unlike PMMA, PC is difficult to thermoform at most forming temperatures, with pressure forming yielding the best products. PC picks up substantial amounts of atmospheric moisture. As a result, the sheet should be kept wrapped in polyethylene until used.

2.5.6 Polyethylene Terephthalate [Polyester, APET, PETG, CPET]

Amorphous polyethylene terephthalate (APET or PETG), is the non-crystalline version of PET, the product used for the very familiar two-liter carbonated beverage bottle. Under certain conditions, PET slowly crystallizes to yield a high temperature, semi-crystalline plastic. With proper extrusion equipment, extruded PET can be cooled quickly enough to prevent substantial crystallinity. The sheet product is called APET, or amorphous PET. If this sheet is quickly heated to its forming temperature and formed against a cool mold, the sheet does not crystallize. Products formed with APET are typically tough, transparent packaging, such as that used with point-of-purchase tool kits or emergency and operating room crash kits.

When long-chain glycols are substituted for some of the polyethylene glycol used to make PET, the resulting copolymer cannot crystallize. One type of copolymer thus produced is called PETG. While these copolymers are more expensive than APET, their thermal stability is desired particularly for thicker sheets.

PET in general is quite moisture-sensitive. The moisture in PET does not manifest itself as moisture bubbles in the final part, but rather the moisture attacks the polymer backbone, degrading it. The result is excessive sag during the heating step and a difficult-to-detect loss in mechanical properties such as impact strength. APET tends to be very tough to trim, necessitating very sharp, heated trim dies. Improper trimming can result in fuzz, angel hair, and substantial trim dust.

The wide use of microwave ovens, both commercially and residentially, led to the development of high-temperature packaging incorporating crystallizable PET (CPET). CPET is basically the same polymer as APET with additives that accelerate the formation of crystallites. When CPET is about 20% crystalline, packaging made from it is capable of withstanding 400 °F (200 °C) oven air temperature for up to one hour. CPET containers are formed by forcing the crystallizing PET sheet against a heated mold. As with APET and PETG, CPET polymer requires substantial drying to prevent moisture absorption which can ruin the formability and mechanical properties of the formed product.

2.5.7 Polyethylene [PE, HDPE, LDPE, LLDPE]

Polyethylene (PE) is the crystalline polymer most often used in thermoforming, because of its very high melt strength, or hot strength. Highdensity polyethylene (HDPE) is the polymer most used in parison blow molding for milk bottles and other containers. HDPE is thermoformed into pallets, dunnage, totes, and many outdoor products. Low-density polyethylene (LDPE) is much softer than HDPE and competes with flexible PVC in many non-transportation applications. Linear lowdensity polyethylene (LLDPE) is tougher than LDPE but softer than HDPE. Polyethylenes are typically formed above their melt temperatures. Care must be taken with LDPE however, as it might sag very quickly in the thermoforming oven.

2.5.8 Polypropylene [PP, PP Copolymer]

Polypropylene (PP) is used because of its excellent high-temperature chemical resistance. PP homopolymer is one of the lowest cost polymers available to thermoformers. It suffers from poor sag resistance and haziness, which minimizes its use in disposable packaging. Because of its poor melt strength, PP homopolymer is typically formed at sheet temperatures just below its melting temperature. As a result, its forming temperature window is quite narrow, perhaps only two or three degrees. Because of this, PP homopolymer is typically pressure-formed.

Ease of forming and polymer cost are both increased when PP is copolymerized with PE. Acrylics in small amounts are also co-reacted with PP to improve the polymer's hot strength, albeit at an increased cost. Additives such as sorbitols alter the crystallizing characteristics of PP to produce a polymer with better melt strength, greater forming window, and better clarity, but at an increased cost. Typically, PP is used only in thin-gauge products. To date, all versions of heavy-gauge PP exhibit excessive sag before the sheet is at forming temperature.

2.5.9 Other Polymers

If a polymer can be produced in sheet form, it can probably be thermoformed. In the olefin family, ethylene vinyl acetate (EVA), a low temperature-melting crystalline polymer, and thermoplastic rubbers, which are blends of natural rubber and either olefins or styrenics, have been successfully thermoformed. In certain instances, specific types of nylons are thermoformed, with prepared meat product packages being typical products. Styrene-maleic anhydride (SMA) is a higher temperature styrenic that can be foamed and formed into products such as fast-food breakfast trays and automobile head liners. Modified polyphenylene oxide (mPPO) is sometimes thermoformed into fire-retardant electronic equipment cases. Polytetrafluoroethylene (PTFE) and its copolymer, fluoroethylene polymer (FEP), find uses as chemically resistant liners. Thermoplastic polyurethane is thermoformed into medical devices such as gloves and condoms. Biaxially oriented polyethylene terephthalate (OPET), is thermoformed into speaker and microphone diaphragms. Polyimide (PI), and polyamide-imide (PAI), are extremely high temperature thermoplastics that are usually cast from solution. Once dried of their solvent, these polymers are thermoformed at very high sheet temperatures into high-performance products such as battery boxes.

2.6 Multilayer Polymers

In the past decade, there have been substantial improvements in coextrusion dies. As a result, multilayer coextruded sheet is finding great applications in thermoforming. Thermoforming often generates significant amounts of trim that for economic reasons, must be reground and recycled. Recycled polymer can often be mixed directly with virgin polymer at the extruder hopper. In certain instances, if the regrind has weaker properties or is discolored, it may be coextruded as the core layer, sandwiched between virgin outer layers. In other examples of coextruded thermoformed products, a specific packaging application may require a two-layer structure. In another application, plumbing fixtures may require cosmetic acrylic surfaces and fire-retardant PVC substrates.

Many outdoor products require the substrate toughness of ABS and ultraviolet or UV-resistant acrylic surfaces. Certain transit vehicles require the substrate toughness of an impact-modified styrenic with an abrasion-resistant fluoroethylene polymer surface. Multilayer rigid packaging is a rapidly growing area in food processing. Frequently, the formable sheet contains a moisture barrier such as an olefin, an oxygen barrier such as an ethylene vinyl alcohol (EVOH), a rigid structure such as a styrenic, and tie layers to "glue" these layers together. Although these structures are tricky to thermoform, the primary limitation is recyclability of the trim. For example, although PVC is process-compatible with many barrier polymers, it cannot be successfully melt-processed in the mixed recycle stream. Polyvinyidene chloride (PVDC) is a well-known oxygen barrier polymer, but it also cannot be melt-processed in the mixed recycle stream.

2.7 Filled Polymers

Most thermoformable polymers are neat, i.e., they contain only polymers and processing aids. There is however, growing interest in forming products with improved modulus or stiffness, requirements that can be met by filled polymers.

Probably the first filled polymer commercially thermoformed was talc-filled PP. As noted earlier, PP homopolymer has poor hot strength and a resulting narrow forming window. When talc is compounded into PP at 20% (wt) or more, the resulting sheet has good sag resistance and an increased forming window. Fillers opacify the polymer however, and although the formed product is substantially stiffer, it has poorer impact strength. Polypropylenes with filler loadings to 40% (wt) are commercially available and special compounds with filler loadings to 60% (wt) have been successfully pressure-formed. As the filler loading increases, the polymers become increasingly difficult to extrude with consistently uniform filler distribution throughout the sheet and without surface irregularities.

PVC is another polymer that is filled and thermoformed. Fillers include mica, talc, and calcium carbonate loadings of 20% (wt). Nylons, such as nylon 6 and nylon 11, have extremely poor melt strengths and so are rarely thermoformed unless filled. Mica and talc are typical fillers for nylons.

Most of the fillers used in plastics are inexpensive inorganics such as calcium carbonate and talc. However many other fillers are used, including iron powders for the products of magnetized parts and aluminum powders for the manufacture of electromagnetic interference (EMI)-resistant, electronic cases.

Although not usually considered as fillers, solid fire retardants such as aluminum trihydrate and antimony oxide are compounded into polymers, and alter the mechanical characteristics of the polymer. For example, certain grades of fire-retarded ABS (FR-ABS) are much stiffer than commercial ABS and, they tend to be more difficult to thermoform into deep mold cavities.

As noted, filled polymers are stiffer at the polymer forming temperature. Furthermore, fillers are inextensible or unstretchable. If filled polymers are stretched too much, microvoids can form in the vicinity of the filler. These voids act as stress concentrators and can cause the end product to fail prematurely.

2.8 Reinforced Polymers

Although long-glass fiber-reinforced polypropylene and nylon were available in sheet form in the 1970s, matched-mold compression molding was the standard way of forming these reinforced polymers. In the 1980s, carbon-fiber reinforced thermoplastics such as polyether ether ketone (PEEK) and polyether imide (PEI) were match-molded into high-performance parts for military aircraft and race car shrouds and canopies. Within the past few years, newer thermoforming machines withcontrolled pressure boxes and special mold designs, have been developed to handle these very-difficult-to-form composites. Polypropylene containing about 35% (wt) long glass fibers find applications in non-cosmetic, structural components in light-duty vehicles such as golf carts and tow-behind trailers. In addition to being very stiff at the polymer melt temperature, glass-reinforced sheet has a tendency to expand or loft as the polymer melts. This expansion, attributed to the method of manufacture, leads to porosity in the product, unless a bladder or pressure mat is used to recompress the composite.

As with fillers, fiber reinforcing elements such as glass and carbon fibers are inextensible. The long fiber length essentially prevents the composite thermoplastic from being stretched, even when the polymer matrix is at or above its normal forming temperature. Excessive forming force can lead to fiber breakage, resin filtration from the fiber structure, and excessive void formation. These factors can dramatically weaken the formed structure.

2.9 Foams

Thermoplastic foams are produced by adding appropriate foaming agents to the polymer during the extrusion process. There are two general categories of thermoplastic foams. High-density foam sheet is extruded by adding a chemical foaming agent to the polymer at the extruder hopper. There are two types of chemical foaming agents. Exothermic foaming agents that generate heat when they decompose are most commonly used with polyolefins and most styrenics. Azodicarbonamide (AZ) is the most widely used exothermic foaming agent. Nitrogen is the foaming gas liberated by AZ decomposition.

Endothermic foaming agents require heat to decompose. Sodium bicarbonate ($NaHCO_3$), commonly called baking soda, is the most widely used endothermic foaming agent. Carbon dioxide and water vapor are the foaming gases released by sodium bicarbonate decomposition. Endothermics are used with styrenics and when the end product is used in contact with food products. Typically, endothermics and exothermics are used together in concentrations of 0.5% (wt) to 1.0% (wt) to produce high-density foams with densities of about 70% of the unfoamed polymer densities. Foamed impact polystyrene (HIPS), polyethylene terephthalate (PET), and polycarbonate (PC) are available as commercial sheet and other polymers can be custom-foamed by many extrusion houses. Typically, high-density foam sheet thermoforms in a manner similar to the unfoamed polymer.

Care must be taken to prevent overheating, because excessive heat causes the foaming gas in the cells to expand. The result is an increasing loss of surface quality. Care must also be taken when pressing the formable sheet against the mold surface. Excessive pressure collapses the cells, resulting in an increase in density and an equivalent decrease in sheet thickness.

One reason for foaming is increased stiffness at the same product weight.. Stiffness is proportional to the modulus of the sheet and the cube of its local thickness. Foaming reduces the modulus in proportion to the square of the sheet density. In other words, if the sheet is foamed 30%, the sheet modulus is about 50% of that of the unfoamed sheet. But, if the part weight is the same, the local sheet thickness increases by 43%, and the cube of the local sheet thickness increases by 429%. Therefore, the stiffness of the formed part increases by slightly more than a factor of two.

Low-density foams are used as thermal insulators and as shock mitigators. Polystyrene dominates low-density foam thermoforming. Polyethylene, crosslinked polypropylene, and PET foams are also thermoformed. Low-density foams are characterized by foam densities of 2 to 10 lb/ft^3 (32 to 160 kg/m^3). These foams are extruded into sheet or plank using modified extrusion equipment.

Although chemical foaming agents are sometimes added to the low-density foam extruder, the common foaming agents are physical foaming agents. Hydrocarbons such as butanes and pentanes, hydrochlorofluorocarbons, (HCFCs or refrigerants), and carbon dioxide, are common foaming agents. These gases are held under pressure in solution in the polymer until the gas-laden melt is pushed through an appropriate shaping die. The gases then come out of solution to form a discrete gas phase. Typically, low-density thermoplastic foams are closed-cell foams. Over time, the foaming gas diffuses out of the foam and air diffuses in. This gas interchange is important in thermoforming, because the internal cell gas pressure increases and the polymer softens during the heating portion of the process, thus increasing the sheet thickness by a factor of two or so. This secondary expansion allows for extrusion of higher-density foam and for increased stiffness in the final product.

Low-density foams, where the density is less than about 20% of the unfoamed polymer density, are more difficult to thermoform than unfoamed polymers. As the foam is heated, the polymer softens and the internal cell gas increases in pressure, thus stretching the polymer membranes. Excessive temperature results in membrane rupture and cell collapse. As a result, thermoplastic foams are usually heated until secondary expansion occurs. The sheet is then formed between matched molds. Low-density polystyrene foams are thermoformed into disposable picnic plates and egg cartons. Low-density polyolefin foams are thermoformed into automotive trunk inner liners and shipping protectors. Relatively deep draws are accomplished primarily by compressing the foam cells during forming. In certain instances, low-density foam is single-side laminated with thin, unfoamed 5 mil (0.005-in or 125 μm) thick sheetstock for improved cut resistance and quality graphics.

3 General Forming Concepts

The simplest thermoforming process consists of simply heating the sheet and forcing it against a solid shape called a mold. There are many variations and improvements of this simple process. This chapter highlights some of the major methods employed in contemporary thermoforming.

3.1 Simple Heating and Stretching

Technically, the basic thermoforming process is one of differential stretching, as shown in Fig. 3.1. Only the sheet that is free of the mold surface stretches. As a result, as stretching continues, the sheet becomes thinner and thinner before it contacts the mold. The areas formed last are the thinnest, the most oriented, and the weakest. As a result, the final thermoformed part is characterized by non-uniform wall thickness.

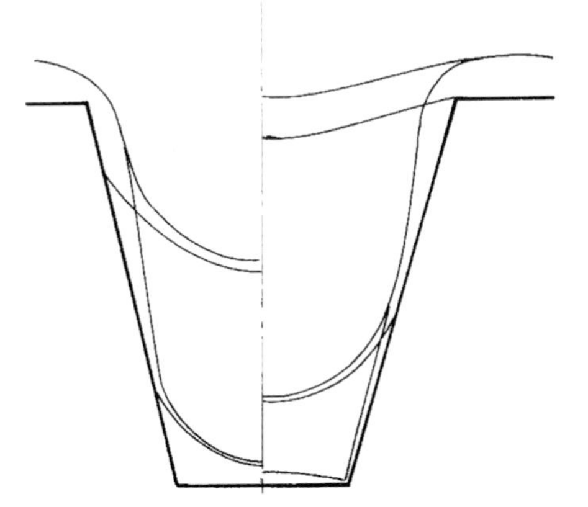

Figure 3.1 Wall thickness variation for simple vacuum forming.

3.1.1 Drape Forming

As shown in Table 1.1, the earliest thermoformed articles were made by simply heating sheet goods until they were soft, and then manually shaping the sheets over simple forms. As shown in Fig. 3.2, drape forming over a male, or positive, mold yields a part that is thinner along its side walls, rim, and corners than at the bottom. Furthermore, the inside of the formed part contacts and replicates the mold surface. Drape forming is used to make such heavy-gauge products as signage and refrigerator liners.

3.1.2 Vacuum Forming

The development of simple vacuum forming in the early part of the 20th century followed the development of simple, electrically-driven vacuum

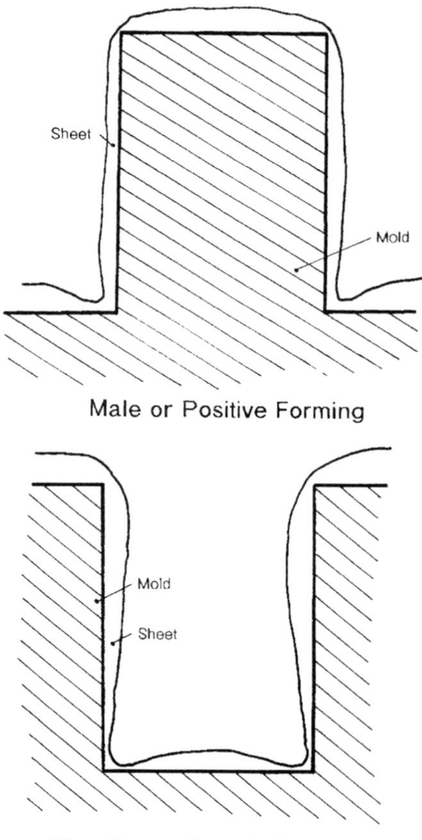

Figure 3.2 Drape forming.

Figure 3.3 Vacuum forming.

systems (Fig. 3.3). As shown in Fig. 3.1, the sheet is stretched into a female, or negative, mold. As a result, the outside of the formed part contacts and replicates the mold surface. A part formed by vacuum forming is thinner in the bottom and corners that at the top or rim, in contrast to drape forming. Vacuum forming is used to make such heavy-gauge products as outdoor signs and such thin-gauge products as picnic plates.

Both vacuum forming and drape forming yield parts with highly uneven wall thicknesses. However, both techniques are commercially used today, primarily for shallow-draw parts, or where wall thickness is not critical to the functioning of the part.

3.1.3 Free Forming

Free forming, also called billow or free bubble forming, is schematically shown in Fig. 3.4. Free forming uses no mold, in contrast to all other forms of thermoforming. Instead, the sheet is heated to its forming temperature, then air pressure is applied against the sheet, and the sheet expands. As the bubble expands, it touches a microswitch or intersects a light beam, which controls the air pressure, which in turn controls the final height of the bubble. Because the softened plastic bubble never contacts a solid surface, it remains mar-free. The bubble is quite uniform in thickness except in the clamping region. Transparent polymers are used most often in this process. Heavy-gauge, free-formed shapes are used as skylights and aircraft windscreens. Thin-gauge, free-formed shapes are used in blister packaging.

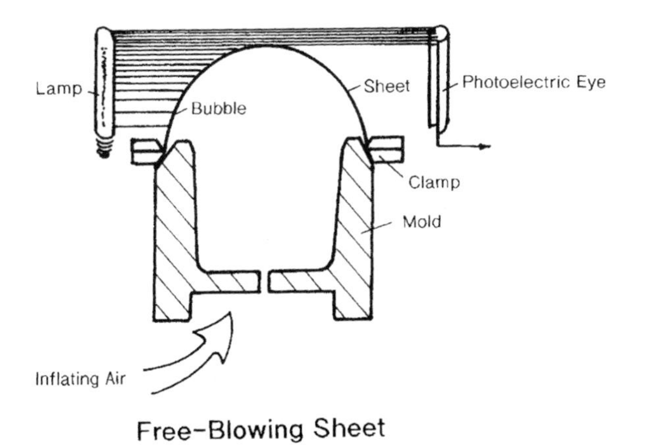

Figure 3.4 Free-blowing.

3.2 Assisted Forming

For all deep-draw parts and most other thermoformed parts, the basic thermoforming processes yield locally unacceptable part wall thicknesses. There are three commercial ways of improving the uniformity of thermoformed parts.

3.2.1 Non-Uniform Heating

For heavy-gauge forming, non-uniform heating, also called pattern, zoned, or zonal heating, produces a sheet that is hotter in certain areas than others. Hotter sheet stretches more than cooler sheet. As a result, regions of the sheet that would normally be overly thinned are not heated as much as regions that would normally be excessively thick. Non-uniform heating is not usually used for thin-gauge thermoforming, since the parts are often much smaller than heavy-gauge parts.

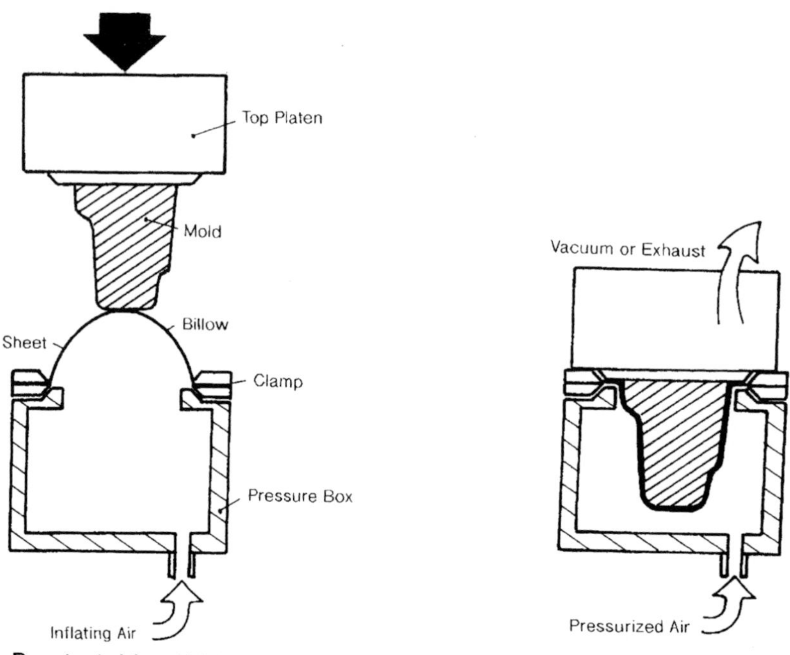

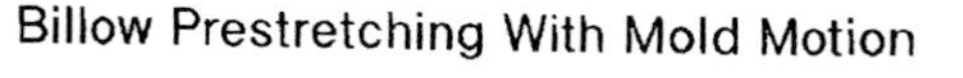

Figure 3.5 Billow prestretching with male mold.

3.2.2 Pneumatic Preforming

In free forming, the sheet has a nearly uniform thickness when it is pneumatically inflated. In several thermoforming methods, this effect is used extensively as the first step to improve the wall thickness uniformity of the final part. One of these methods, called billow drape forming, is schematically shown in Fig. 3.5. The sheet forms around the mold as the mold is plunged into the inflated sheet. Vacuum is applied to the mold to ensure that the sheet replicates the mold surface. In another version, the male mold is raised into the inflated sheet.

Draw-box preforming is another way of prestretching the sheet (Fig. 3.6). In this case, the heated sheet is drawn with vacuum into a basically empty box, called a draw box. The male mold is then plunged or immersed in the stretched sheet. The draw-box method is preferred when the polymer may be difficult to stretch without localized blow-out. Rigid PVC and fire retarded-ABS are polymers that are successfully prestretched using the draw-box method.

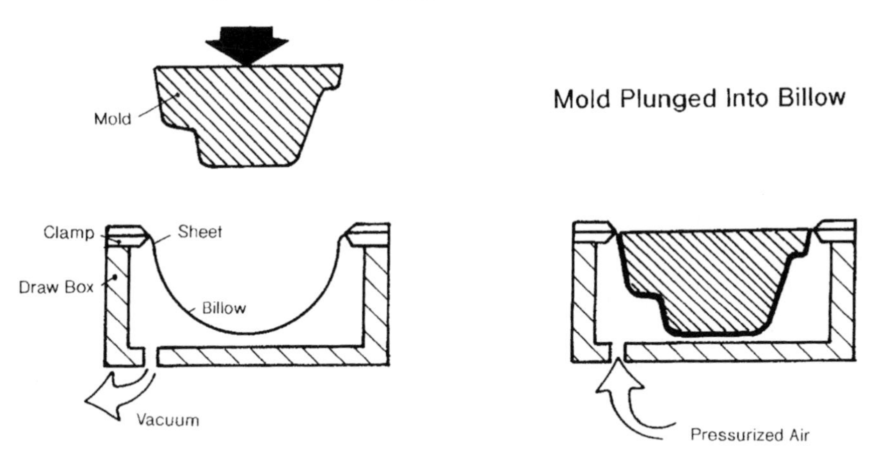

Figure 3.6 Vacuum draw-box with male mold.

3.2.3 Plug Assist

Plugs, also called assists or pushers, are mechanically driven, shaped solid structures that are pressed into the softened sheet prior to forming. Plugs are primarily used to locally stretch a sheet. Figure 3.7 shows a schematic of the

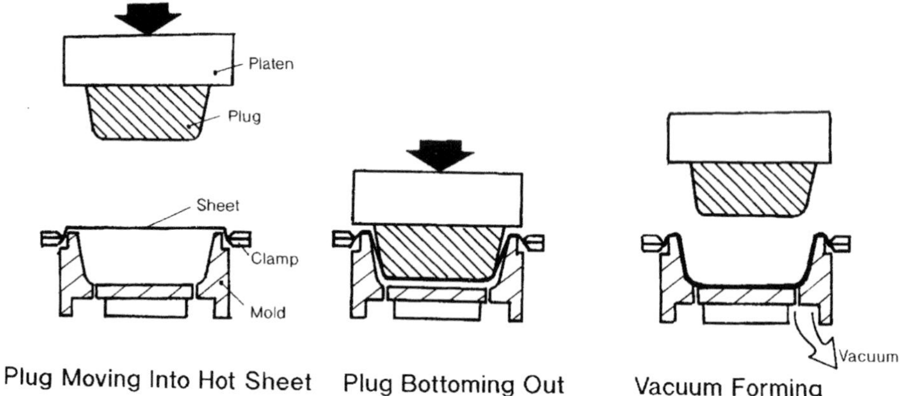

Figure 3.7 Plug assist with female mold.

simplest and most common form of plug-assisted thermoforming. As discussed in Chapter 8, plugs can be made of any heat-resistant solid material, including wood, plywood, medium-density fiberboard (MDF), syntactic thermosetting foam, plastics such as nylon and FEP, and heated aluminum. The choice of plug material and the shape of the plug depends on the type of polymer to be formed and the shape of the final part. Plugs are used extensively in thin-gauge forming for products such as drink cups, and in heavy-gauge forming for products such as tote bins and equipment cabinets. Because plugs are solid surfaces that are cooler than the softened sheet, they tend to "mark off" or leave their impressions on the final formed product.

3.3 Pressure Forming

Technically, all thermoforming methods except mechanical bending and shaping, employ differential pressure to stretch the sheet against the mold surface. Therefore, all thermoforming can be considered pressure forming. However, according to accepted terminology, thermoforming is considered to be pressure forming only when the differential pressure across the sheet thickness exceeds one atmosphere (15 lb/in^2 or 0.1 MPa).

Traditional pressure forming uses air pressure of up to ten atmospheres (150 lb/in^2 or 1 MPa) on the free side of the sheet and vacuum on the sheet surface closest to the mold. The air pressure is contained in a pressure box that clamps the sheet against the mold surface. Pressure forming is used in heavy-gauge forming when the sheet, at forming temperature, is too stiff to

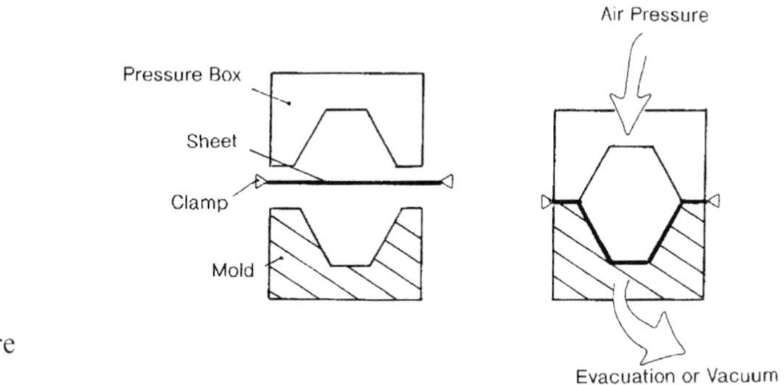

Figure 3.8 Pressure forming.

adequately replicate the mold surface. Pressure-formed parts have surface textures and radii that rival injection molded parts. Pressure forming is used in thin-gauge forming to improve cooling cycle times by rapidly stripping the sheet from the plug and driving it against the cold mold. A simple schematic of pressure forming is given in Fig. 3.8.

Heavy-gauge pressure forming is also employed with the traditionally stiffer filled polymers. Two-sided molds, sometimes called matched molds, are also used to form filled polymers, where pressures of ten atmospheres (150 lb/in^2 or 1 MPa) are employed. Foamed sheet is also molded between

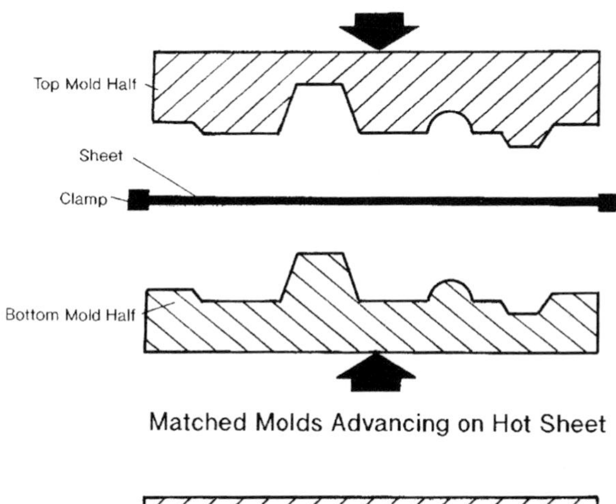

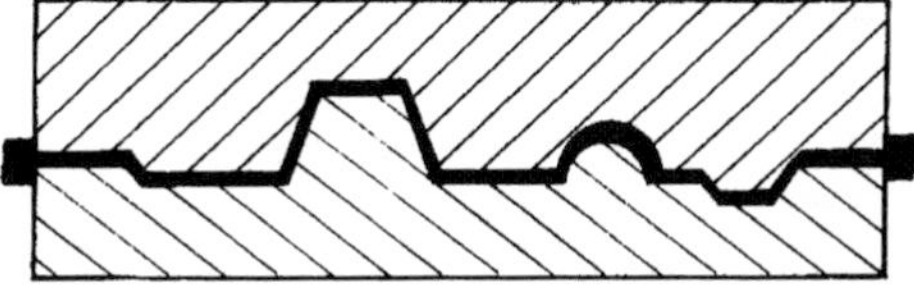

Figure 3.9 Matched mold forming.

two mold halves (Fig. 3.9) at pressures of about three atmospheres (45 lb/in^2 or 0.3 MPa). Pressures of ten atmospheres (150 lb/in^2 or 1 MPa) are also used for short-glass fiber-reinforced polymers. Recent work involving one-sided molds and pressure bladders over the free surface of the forming sheet, similar to thermoset composite forming, holds promise. Higher pressures, approaching compression molding pressures of 100 atmospheres (1500 lb/in^2 or 10 MPa), are needed to shape long-glass and continuous glass fiber-reinforced polymers.

3.4 Twin-Sheet Forming

Although the practice of heavy-gauge, twin-sheet thermoforming is decades old, recent technical advances in machines and molds have made it commercially competitive with rotational molding and blow molding. There are at least three commercial methods used today. In the oldest method, used in heavy-gauge forming, the two halves of the end product are manufactured independently on simple, single-sided molds, then glued or thermally welded together (Fig. 3.10).

In simultaneous twin-sheet forming, both sheets are clamped in a single frame with a blow pin between them (Fig. 3.11). During heating, the space between the sheets is pressurized enough to keep the sheets apart. The sheets are then formed in a two-sided mold. The top sheet is formed into the top mold half and the bottom sheet is formed into the lower mold half. In sequential twin-sheet forming, the first sheet is heated and formed into the lower mold half, and the second sheet is then formed into the upper mold

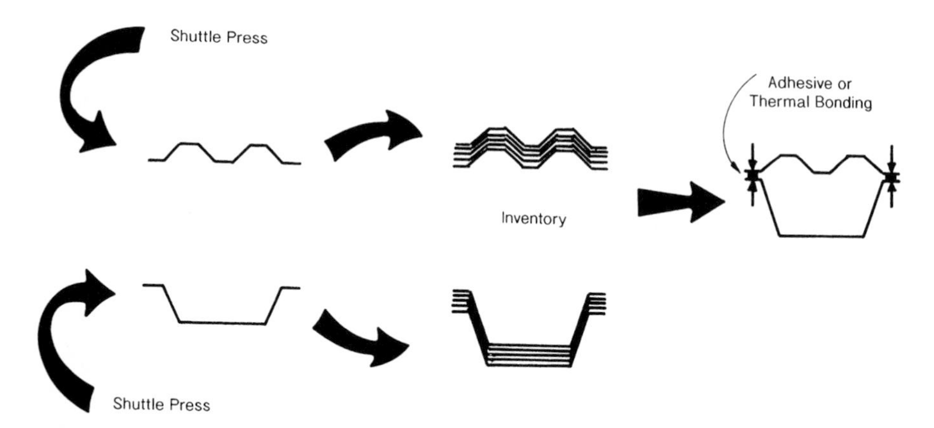

Figure 3.10 Twin-sheet production using inventoried method.

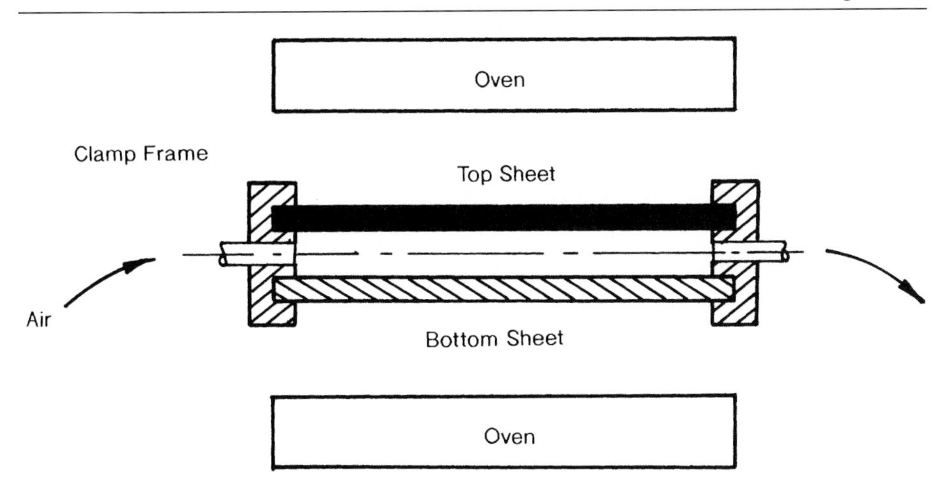

Figure 3.11 Simultaneous twin-sheet production.

half. The two mold halves are then brought together to form the hollow or semi-hollow part (Fig. 3.12).

Thin-gauge twin-sheet forming was practiced several decades ago, to manufacture products as diverse as ping pong balls and narrow-necked bottles. Recent technical advances in temperature control and mold design have enabled very high speed twin-sheet forming of narrow-necked containers for dairy products (Fig. 3.13). Two sheets are independently heated, either with non-contact infrared heaters or by direct contact heaters, and then brought together in the forming press. Special, high-speed trimming presses, similar to those used in blow molding, separate the containers from the web.

3.5 Contact Forming

It is not economical to heat very thin plastic sheets by non-contact infrared means. The polymer may be too transparent to infrared energy, may cool too quickly because of its high surface-to-volume ratio, may sag excessively unless supported, or may be thermally sensitive. As a result, direct contact heating, sometimes called trapped sheet heating, is used. The sheet is brought in contact with a heated, non-stick metal surface, then quickly transferred to the mold for forming. The sheet can contact either one heated surface (Fig. 3.14) or two surfaces (Fig. 3.15). Contact heating is frequently used in conjunction with form-fill-and-seal methods, where the final

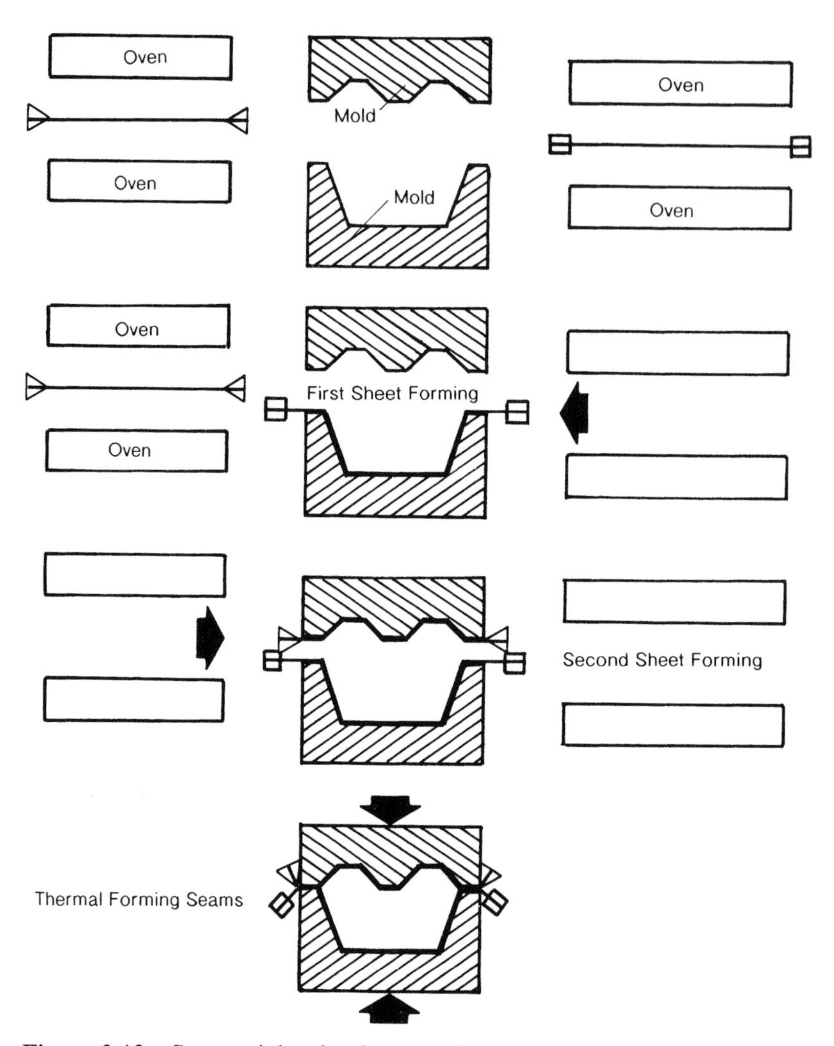

Figure 3.12 Sequential twin-sheet production.

product is a sealed, rigid container containing a foodstuff, a medical product, or a hardware product.

3.6 Diaphragm Forming

Polymers that tend to split or are weak at forming temperatures pose special forming problems. Diaphragm forming or bladder forming is an alternative

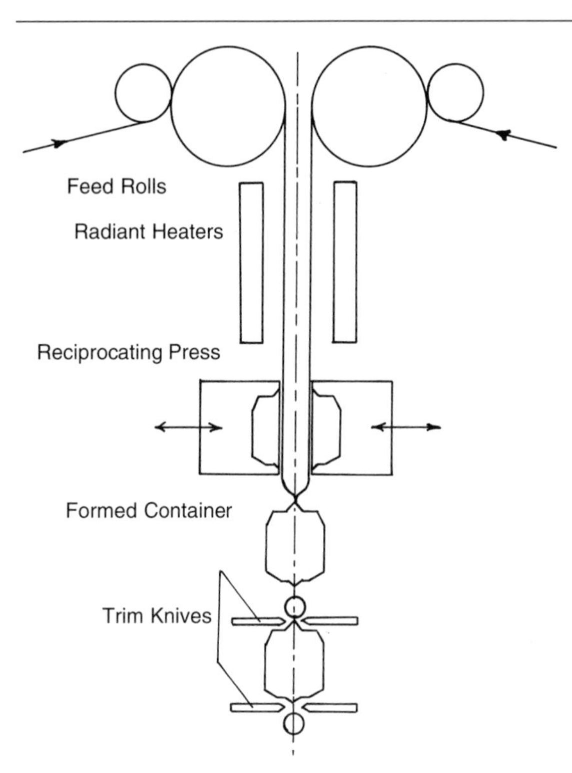

Figure 3.13 Thin-gauge twin-sheet thermoforming

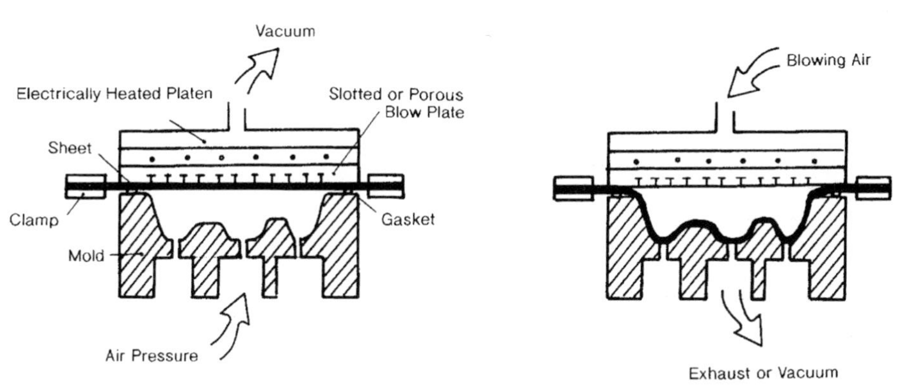

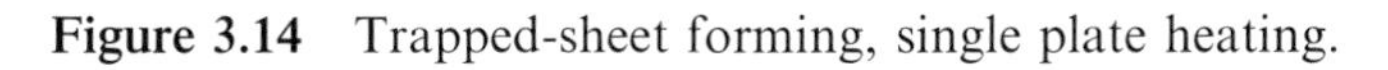
Figure 3.14 Trapped-sheet forming, single plate heating.

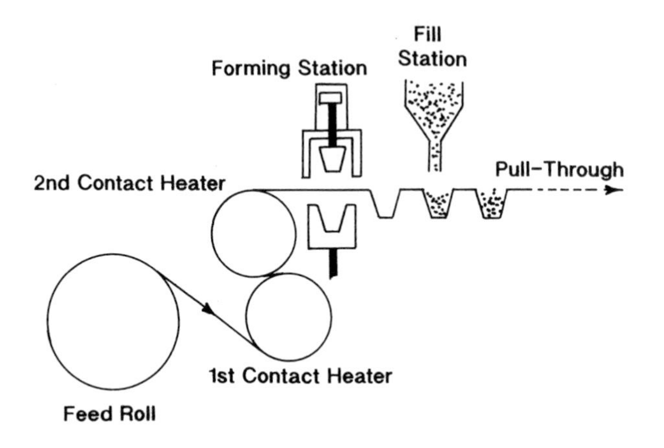

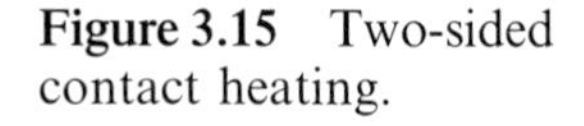
Figure 3.15 Two-sided contact heating.

for thin-gauge sheets (Fig. 3.16). The diaphragm is usually a high-temperature rubber such as neoprene. The sheet is clamped against the diaphragm and heated. The diaphragm and its hot sheet are then deformed against the mold surface, using hydraulic oil or hot water as the forcing fluid. Part wall thickness is quite uniform. With proper conditioning, the diaphragms are useful for hundreds of cycles

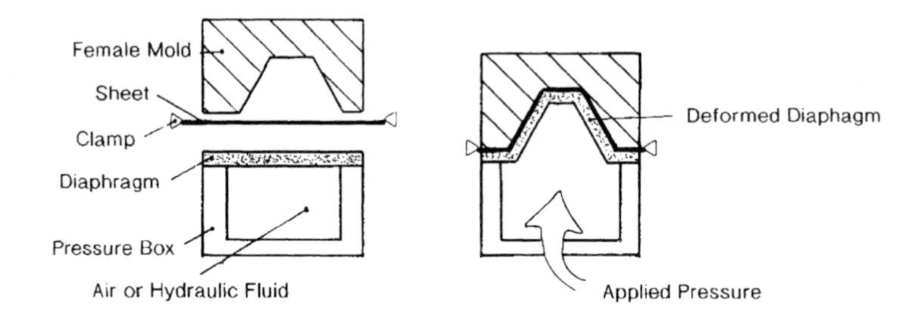

Mold Moving Onto Diaphragm/Sheet

Diaphragm-Stretched Sheet

Figure 3.16 Diaphragm forming.

3.7 Mechanical Forming

As mentioned earlier, hot sheet can be manually stretched against mold surfaces. For simple bends, strip heaters are placed against sheet held in a bending fixture. When the local area of the sheet is hot, the bending fixture is activated and the sheet is allowed to cool in its final, bent form. Barrel skylights, characterized as partial cylinders, are made by fixturing one edge of a sheet in one edge of a cylindrical fixture, then heating the sheet until it sags against the surface of the fixture. The free end of the sheet is then captured in the other edge of the fixture and held until the sheet retains the shape of the fixture. Care must be taken to ensure that the plastic is sufficiently heated. Otherwise some spring-back occurs, and the initially molded angle gradually opens.

4 Machinery for the Forming Process

In 1996, the US thermoforming machinery market was estimated to be about US $300 million, not counting the rebuilding and upgrading markets. Because thermoforming is typically a relatively low temperature, low pressure process, many thermoforming machines are home-built. It is estimated that about 40% of the currently operating thermoforming machines are either home-built or substantially modified from the original designs. Of course, many thermoforming machines are built for specific purposes or have limited, non-critical use. As an example, a thermoforming machine to manufacture simple shallow-draw, thin-gauge signs might consist of a wooden hinged book mold to hold the sheet, a bank of sun lamps, a vacuum box with a fine welded wire screen on which block letters are assembled, and an inexpensive shop vacuum. Another machine to manufacture simple display signs might consist of a surplus domestic oven and a wooden mold. This chapter focuses on commercially manufactured machines designed to produce quality parts with high productivity, high reliability, and low maintenance costs.

4.1 Thin-Gauge, Roll-Fed Machines

Figure 4.1 is a schematic of a relatively common roll-fed thermoforming machine for the production of thin-gauge parts. The functions of the machine, beginning with the take-off roll stand, are detailed in the figure. Salient features of each of these functions are described below.

4.1.1 General Features

Several defining characteristics should be considered when choosing a roll-fed machine. They include platen dimension, maximum depth of draw, the nature of the forming process, the types of power drive for the platen and for sheet indexing, the type of heater recommended by the machinery

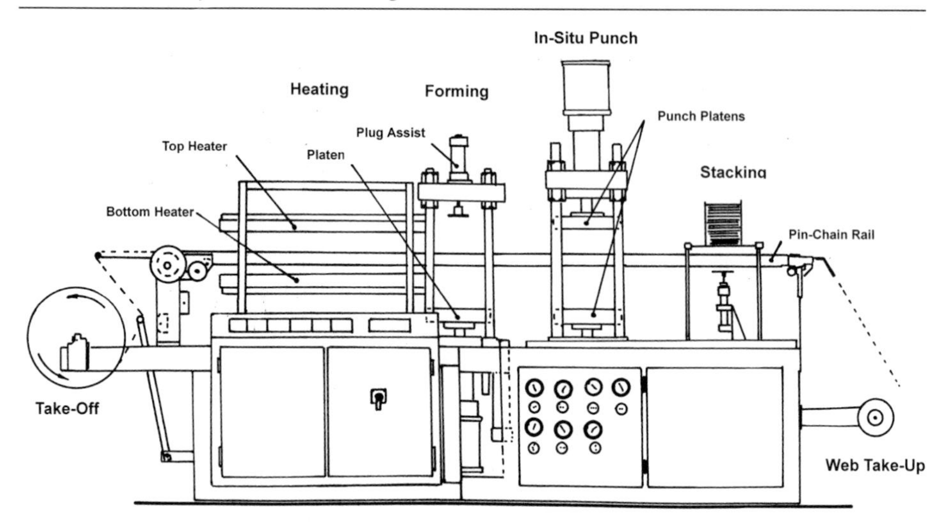

Figure 4.1 Thin-gauge roll-fed thermoformer (Kiefel).

builder, the type of heater process control, and recommended maximum heater output, in W/in^2 or W/ft^2. Other features, such as form-fill-and-seal, trim-in-place, in-press stacking, all-electric drives, and process control options should also be considered when discussing machine purchases with the machinery builders.

4.1.2 Sheet Take-Off

Incoming sheet is delivered as rolls for the stand-alone, thin-gauge forming machine. The roll stand should be capable of handling 2000 lb (900 kg rolls) or rolls of up to 6 ft (2 m) in diameter. It should accommodate rolls wound on cores with different diameters, from 3 in (75 mm) to as large as 8 in (200 mm). The roll stand must have passive braking or a roll speed governor and an end-of-roll alarm. Rapid roll changeover is an optional feature.

4.1.3 Pin-Chain and Pin-Chain-Rail

Typically, thin-gauge sheet is advanced through the machine on pins spaced along the lengths of parallel or near-parallel, continuous bicycle-link chains. The pins should be removable for sharpening or replacement. Typically, the pin-chain-rail assembly is shielded from the oven heat. For certain polymers such as PET and RPVC, pins are heated to minimize sheet splitting. Self-lubricating chains have advantages, but chains that run dry or without lubricants are required for medical and food applications. Lower pin guides are needed to keep the pins vertical during heating and forming. In addition, options might include chip vacuum at the pin-sheet engagement, non-stick or non-scratch engagement shoes just ahead of the pin-sheet engagement point, an "out-of-sheet" klaxon alarm, and automatic parallel chain adjustment. Servo-driven chain advancement is usually standard and offers smooth and constant sheet acceleration and deceleration rates and constant sheet speed during advancement.

4.1.4 Oven

Multicavity or family molds are commonly used in thin-gauge thermoforming. As a result, it is often necessary to provide as uniform energy input to the sheet as possible. Many thermoforming machines have modular ovens to allow more than one "shot" to reside in the oven. Two or three oven sections are recommended for hard-to-heat polymers such as APET, PVC and PP. As many as five oven sections are used for low density styrenic and olefinic foams. Preheating ovens, consisting of a series of up-and-down roller-driven loops, are recommended for PP. The preheating oven is placed between the take-off roll stand and the pin-chain-rail portion of the machine. Ovens should be well-insulated and the sides should come to within 1 in (25 mm) of the machine framework that holds the chain-rails. Baffles are sometimes used between oven units and between the last oven unit and the mold.

Some manufacturers provide a means of changing the sheet-to-heater gap on both top and bottom heaters. Servo-driven gap adjustment is a useful option. Provisions should be made to quickly retract the oven whenever the chain travel is interrupted. Mechanical means include horizontal pull-away and clam-shell fly-open. Rapid disconnects should be provided for individual heater elements and thermocouples. Most machinery builders provide openings in both top and bottom oven surfaces for infrared temperature measuring devices. A photo-eye sensing and warning system should be specified when running PP or other saggy polymers. Sag bands are PTFE-coated rods or continuous wires that are parallel to and between the pin-chain rails. They support the sheet in the last portion of the oven. Carbon dioxide fire extinguishers are options when running polyolefins. There should also be a means for accessing the heaters for inspection, repair, and replacement. The forming and cooling steps are usually the controlling steps in thin-gauge thermoforming. As a result, the sheet exits the oven based on time spent in the oven rather than based on the sheet's surface temperature.

4.1.5 Press

Once the sheet is hot, it must be quickly pressed against the mold's surface. Experience shows that it is not uncommon for thin-gauge sheet to lose 20 °F (10 °C) or more in the time between its exit from the oven and the closing of the forming press.

The press has many functions. It must close smoothly and rapidly against the sheet, without banging the mold sections. Some press clamping is required. Pneumatic and servo-mechanical linkages are standard. Platen locking devices and pneumatic glands are used when increased forces are required in pressure forming. The all-electric servo-drives allow for similar locking forces. Platen screws should be either self-lubricating or continuously lubricated, although unlubricated screws are required for medical and food applications. In addition, self-leveling platens configured to easily accept mold changeover are desired. When in-place trimming is sought, platen locking cogs and screws must be protected from trim dust, chips, dirt, and other detritus. There must be adequate daylight between platens to allow inspection and maintenance and for replacement of in-mold trimming dies.

Modern presses are designed with adequate space for such features as plug assist plates, cavity isolator plates, trim-in-place die plates, and ejector ring plates. Water lines and vacuum lines must be properly located and easy to disconnect. Most machinery builders can provide features such as rapid and accurate platen alignment, mold alignment, and ancillary device alignment. Pneumatic interlocks prevent premature air pressurization for pressure forming, premature pressure box opening, and pressurization of an empty mold. PLC controls should allow adjustment of all rate-dependent ancillaries such as plug assist motion, cavity isolator motion, pressure box motion, and trim-in-place sequencing. Both upper and lower platens should be capable of moving across the pin-chain plane for easy mold installation, mold removal, and adjustment of ancillary features.

4.1.6 Forming Assist Devices

Modern thin-gauge thermoforming frequently requires cavities to be isolated from one another. Plug assist and pressure forming are common, particularly for deeply drawn parts such as drink cups. Plug assist features should include capability for the rapid replacement of individual plugs, for internally heating and/or cooling aluminum plugs, for relatively easy removal of the entire plug assist platen, and for relatively easy adjustment of plug travel and rate of travel. The cavity isolator plate and the ejector plate should be easily removed as well.

4.1.7 Trim Means

In Chapter 7, are further details about three general ways of trimming thin-gauge parts from web. In in-line trimming, the sheet with the formed parts still attached to the web is conveyed away from the thermoforming machine to a separate, in-line, trimming press. Figure 4.2 shows the combination of a forming machine, followed by an in-line canopy trim press. In in-machine trimming, the sheet with the formed parts still attached to the web is conveyed by pin-chain from the forming press to a separate in-machine trimming station that may also contain a stacking feature. In in-place trimming, the trim die is an integral part of the forming mold and may also act as the cavity isolator. Usually in in-place trimming, the parts remain tab-attached to the web, with punch-out separation in a separate, in-machine stacking station. For in-place trimming, there should be a rapid way of determining the sharpness of the individual trim dies and a means of quickly adjusting individual trim dies for concentricity and parallelism, as

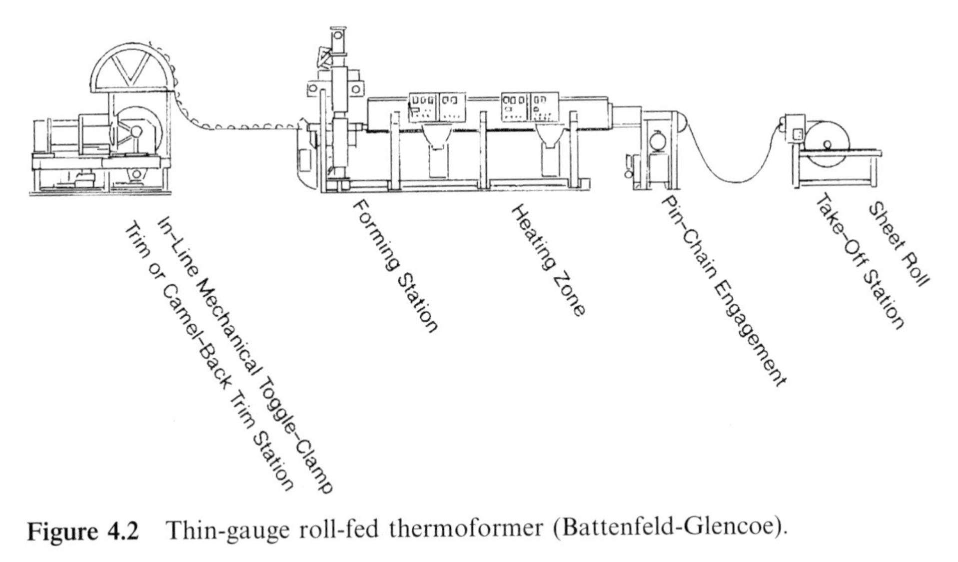

Figure 4.2 Thin-gauge roll-fed thermoformer (Battenfeld-Glencoe).

well as an easy way of removing a given trim die. Automatic chip, trim dust, and detritus removal are desired for brittle polymers such as GPPS.

4.1.8 In-Machine Stacking Means

Many machines include stacking means (Fig. 4.1). The parts are punch-separated from the web using either mechanical piston linkage or rack-and-pinion servo drives. In dedicated machines or machines designed to produce only one product, such as drink cups, yogurt cups, or deli containers, the parts may be removed from the forming mold by vacuum suction cups. Rotating the mold to dump the formed parts is another technique gaining interest.

4.1.9 Trim or Web Take-Up Station

There are two general ways of treating the web, trim, or skeleton. One is to guillotine the web directly below the pin-chain and to feed the guillotined sections directly into a grinder situated below or near the guillotine. Another is to continuously wrap the web into a roll. The take-up station must include a passive tension brake and may include a pressure or hold-down roll. The web roll tends to be very bulky and difficult to handle.

4.1.10 Condition Monitors

Most thermoforming machinery builders provide optional through-the-oven-wall infrared thermometers. Because the forming, cooling, and in-machine trimming steps usually dictate the total time step, the temperature measurements are best used during process set-up. Mold and coolant temperatures should be monitored at one and preferably, two locations in a multicavity mold. Heater temperatures should be monitored at as many places as possible with readouts displayed in false color on the PLC.

Air pressure and vacuum pressure read-outs should also be available on the process computer. Because the cycle step times for thin-gauge forming are tens of seconds or less, very accurate electronic timers are preferred over mechanical clock timers. Since time steps in various sequences can be only fractions of seconds apart, air valves that respond very rapidly are needed. This is critical for the sequenced activation of cavity isolator platen, plug assist platen, pressure box, ejector ring platen, and trim-die platen, particularly during cut-through. This is true also for vacuum steps, including cavity evacuation. A pressure monitor should be provided on the pressure box, and vacuum monitors are needed on the vacuum pump, surge tank, vacuum box, and at least one cavity. Photoelectric eyes and alarms are recommended in the oven for excessive sheet sag and "out-of-sheet" indication. In certain instances, "part in mold" indication is necessary.

4.1.11 Process Control

Most machines come equipped with PLC controllers and color monitors. Again, although sheet temperature can and should be measured, nearly all machines exit the sheet from the oven to the forming station based on clock time. Again, electronic timers are desired because the stepping times are frequently in seconds and fractions of seconds. Most important are automatic protocols for emergencies such as fire, power overload, brownout, power outage, light curtain interruption, and safety cage breach. Not only should automatic actions be taken, such as pulling heaters away from sheet, but klaxons should sound and the nature of the interruption recorded electronically for future review. Computers allow for extensive data logging, including start-up times, heater configuration, hourly energy consumption, actual sheet temperatures, and power and cycle interruptions.

4.2 Heavy-Gauge Sheet Forming

There are two general types of machines used to form heavy-gauge sheet. More than 70% of the commercial heavy-gauge thermoforming presses in use today are shuttle presses (Fig. 4.3). Although shuttle presses are very versatile and capable of forming parts of nearly unlimited dimension, they are economically inefficient. Rotary presses, either three-station (Fig. 4.4) or four-station (Fig. 4.5), are quite energy efficient, but require more care in setting up and are limited in the sizes of parts that can be formed. As with roll-fed machines, machine criteria include platen dimensions and depth of draw; the general nature of the forming process such as vacuum, pressure, matched mold, plug assist, and twin-sheet capacity; the types of motive power for moving the platen and indexing; the manufacturer's recommended types of heaters; heater control; and maximum energy output in W/in^2 or W/ft^2. In addition, many other features, such as automatic oven-to-heater adjustment, deep draw capability, drop-sided ovens, vacuum

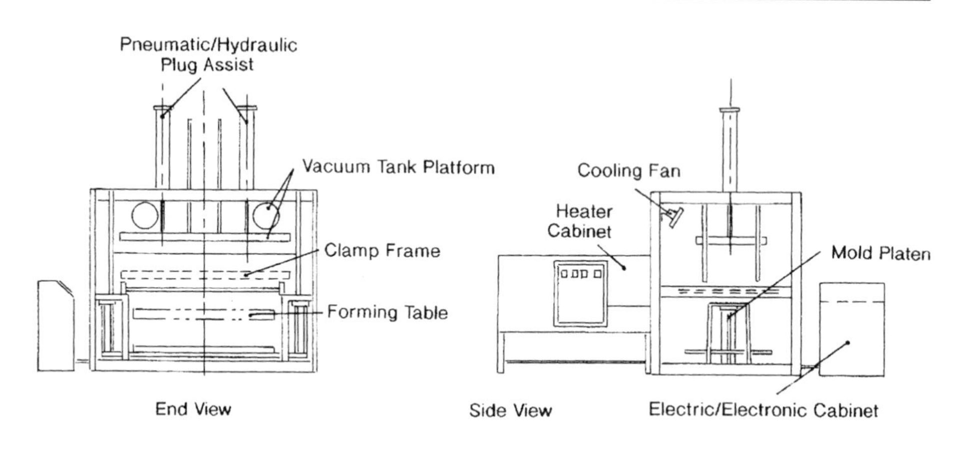

Figure 4.3 Heavy-gauge shuttle thermoformer (Drypoll).

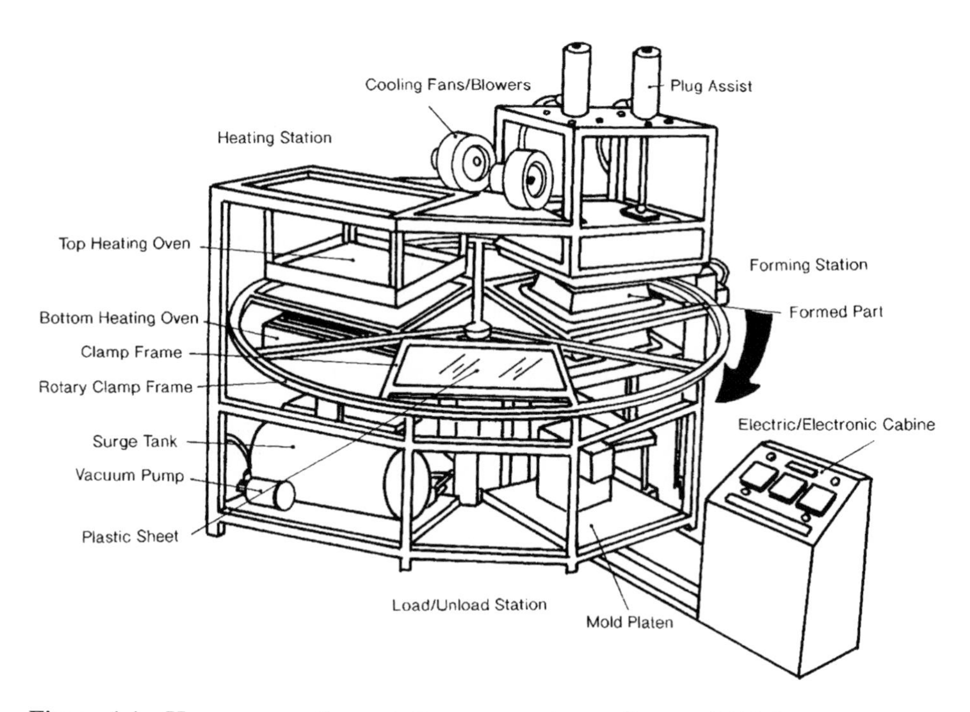

Figure 4.4 Heavy-gauge three-station rotary thermoformer (CAM).

draw boxes, microprocessor controls, automatic sheet loading, through-oven temperature monitoring, and sag monitoring are available as options. The salient features common to cut-sheet or heavy-gauge thermoforming machines are detailed below.

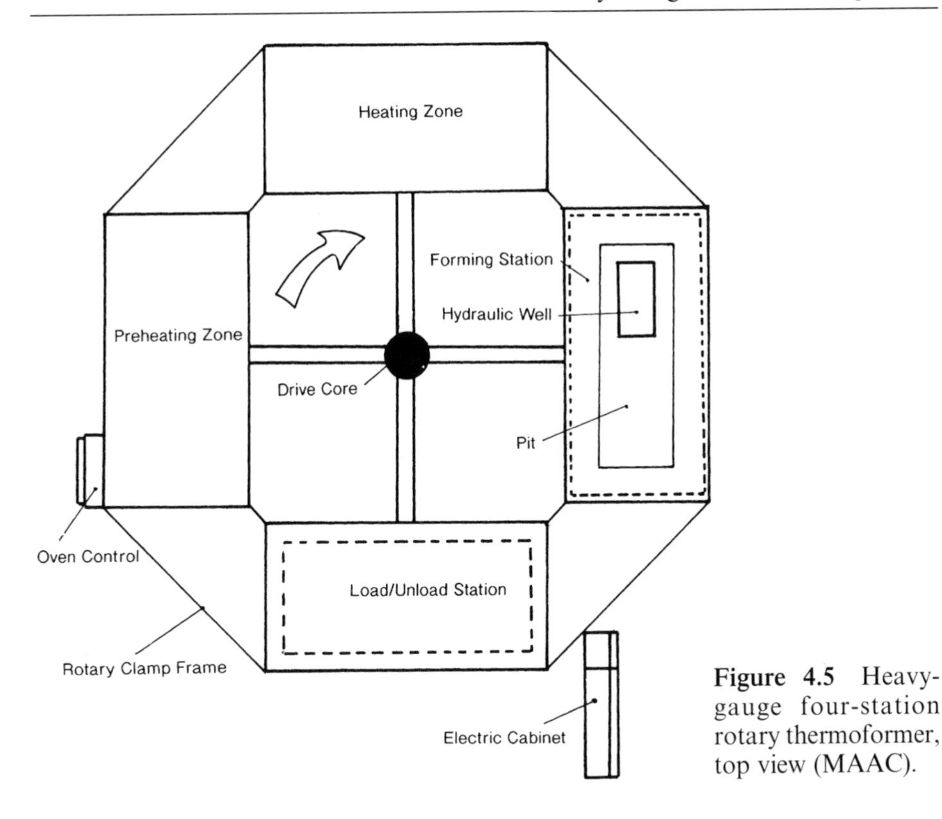

Figure 4.5 Heavy-gauge four-station rotary thermoformer, top view (MAAC).

4.2.1 Shuttle Press

The simplest thermoforming press is a shuttle press. The sheet is clamped in a four-sided clamp frame. The sheet and the clamp frame are then moved into the oven. When the sheet is at its forming temperature, the clamp frame and the sheet are moved from the oven to the forming station. The male mold is moved into the clamped sheet, or the sheet is drawn into the female mold. The formed part is held against the mold until it is cool. Then the clamp frame is removed and the formed part with its excess material to be trimmed off still attached is removed from the mold. The formed part is then separated from the excess material in a secondary trimming operation.

In this simple press, no other sheet is formed while a sheet is heated. Similarly, while a sheet is being formed, no other sheet is heated. There are two ways of circumventing this inefficiency. For the double-oven shuttle press shown in Fig. 4.6, one sheet is formed while a second is heated. This press design is also used to produce twin-sheet formed parts.

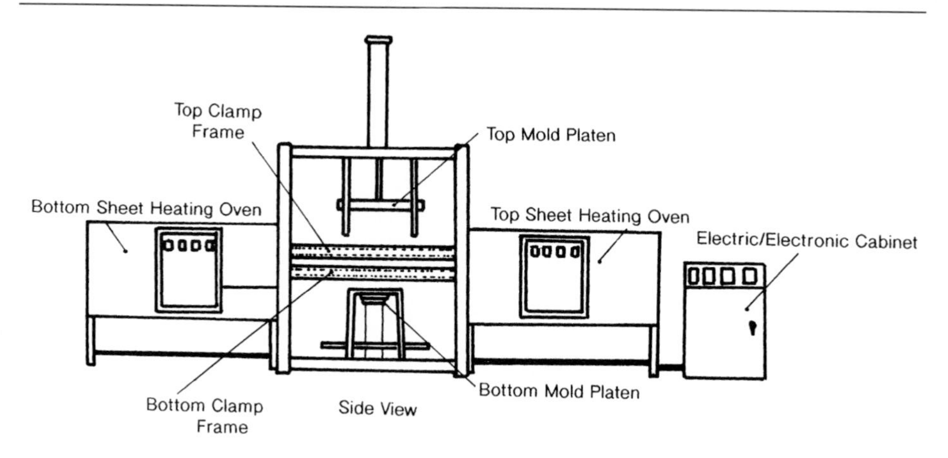

Figure 4.6 Heavy-gauge double-oven shuttle thermoformer.

Another technique uses a rail system similar to that used in roll-fed, thin-gauge thermoforming. Mechanically actuated clamps hold the sheet instead of pins. The chains convey cut sheet into the oven, then retract to pick up another sheet. When the sheet is at the forming temperature, the chains index it into the forming station. They then retract to bring the next sheet from the oven to the forming station. The cabinet shuttle press design common in Europe shuttles the sandwich heaters rather than the sheet. The sheet is positioned over a cabinet that contains the mold. When the sheet is heated to its forming temperature, the oven is retracted and the mold is raised through the sheet clamp plane. This technique works best for male drape-formed parts, but it is also used with plug-assisted female molds.

4.2.2 Rotary Thermoformers

The heart of the rotary press is the carrousel that carries the sheet from station to station. The three stations on the machine shown in Fig. 4.4 consist of a load/unload station, an oven station, and a forming press station. A second oven station is added for the four-station machine shown in Fig. 4.5. Four-station machines are used when sheet heating controls the process cycle or when sequential twin-sheet thermoforming is practiced. The sheet is clamped on all sides at the load/unload station. It is then indexed into the oven or ovens by the rotation of the center turret, which is servo- or hydromechanically driven. The heated sheet is then indexed into the forming press. The formed part, still clamped in its frame, is then rotated from the forming station to the load/unload station, where it is removed, a new sheet clamped in the frame, and the process repeated. One part is molded for each index of the carrousel. Indexing time is dictated by the slowest step.

4.2.3 Sheet Handling

When sheet of standard dimensions is used or when production runs are long, sheet is picked and placed mechanically. Commonly, sheets are either lifted with vacuum suction cups or elevated with pneumatic lifting tables. Regardless of how many vacuum suction cups are used, the cup sizes should be sufficient to allow the sheet to be held by only two cups. Vacuum break must be properly controlled to allow the smooth release of the sheet after the clamps are engaged. A center support is sometimes needed beneath the clamp frame when very large sheets are manually loaded.

4.2.4 Sheet Clamp

If a hinged, book-mold clamp frame is used, the frame must have adequate lock-over toggle clamps. When pneumatic clamps are used, the clamp frame should have barbs on the closing portion and the fixed portion should be either smooth or have a continuous lip. Clamp pressure should be at least 50 lb/in^2 (0.33 MPa) against a cold sheet to minimize extrusion of the heated sheet from the clamp frame. The clamping area should hold at least 0.5 in (12 mm) of 0.100 in (2.5 mm) thick sheet and at least 2 in (50 mm) of 0.400 in (10 mm) thick sheet. The clamp frame, pneumatic closure devices, hoses, rotary hose connections, and all electrical leads must withstand at least 800 °F (425 °C) for 20 minutes for at least 10,000 cycles without sticking, binding, or leaking fluids.

Self-lubricating hinging action to minimize sheet contamination is desired. The clamp frame must be quickly adjustable for various sheet dimensions. For rotary machines, the carrousel must be capable of supporting the maximum sheet weight without oscillation and bouncing, and rotation acceleration, constant speed, and deceleration must be smooth, without vibration. Most machines use a drop pin arrangement to ensure that the carrousel is positioned positively at each operating station.

4.2.5 Oven

For rotary forming presses, a preheat oven is recommended for hygroscopic polymers such as ABS and acrylics. It is also required for sequential or "drop-sheet" twin-sheet thermoforming. For shuttle machines with a single oven, the oven temperature should be step-controlled. Side baffles or oven side walls should close off the sheet and clamp frame during the heating step to minimize drafts and optimize energy usage. As with roll-fed ovens, rapid disconnects for individual heater elements and individual heater thermocouples are desired. Ovens should be provided with ports for in-oven infrared temperature sensing devices. For ovens with very large areas or when polymers with a great deal of sag are processed, the lower portion of the oven should have a drop side toward the press station. Some machine designs allow the lower oven to tilt toward the press station.

As with all thermoforming ovens, provision should be made for emergency shut-down. Baffles that automatically close between the sheet and the heaters and air blown at high velocities across the sheet surfaces are two ways of preventing the sheet from overheating. In addition, photo-electric eyes to monitor sag and carbon dioxide fire extinguishers are desired features.

The spacing between the sheet and the top and bottom oven halves should be easily adjustable. At its maximum, the spacing should be sufficient to allow manual adjustment of individual heaters, heater temperature measurement, burn-out inspection, and element replacement. Machine manufacturers may offer models that allow the bottom oven to lower during sheet heating as the sheet sags, quartz plates over the lower heaters to prevent damage to the heaters if a sheet drops, and even intermittent vacuum or air lift of the sheet to minimize sag.

4.2.6 Press

Adequate press capacity is probably the most critical part of heavy-gauge machine design. Press clamp frames must be sufficiently robust to allow molds to be affixed over the sheet as well as placed beneath the sheet. Other features requiring sufficient machine strength include pressure boxes, plug assist platens, and trim-in-place, forged trim dies. Modern machines are designed with smooth-acting, constant velocity, pressure closure capabilities, as well as smooth acceleration and deceleration at stroke ends to minimize mold banging and chatter as it enters the sheet.

All overhead air and oil lines should be enclosed or self-sealing to minimize contamination of the sheet. All platen locking cogs and screws should be protected from dust and detritus. The press itself should have sites for platen leveling; rapid platen alignment, both vertically and horizontally; rapid vacuum, pressure, and cooling line disconnects for rapid mold changeover. The press should contain pneumatic interlocks to prevent premature air pressurization before the pressure box is locked onto the mold and premature disengagement of the pressure box while it is still pressurized. Locking toggles and air bladders are commonly used to ensure that the mold and pressure box remain secured during pressure forming.

Although the timing sequence is not as critical as with thin-gauge thermoforming, electronic timers are now used to activate and order the various time-dependent functions such as plug assist movement, pressure box engagement and disengagement, vacuum application, blow-back to release the part from the mold, and even trim-in-place devices. Mold changeover tends to be labor-intensive for heavy-gauge formers, and may be even more difficult if some process adjustments must be made below the plane of the shop floor. In certain instances, operators must don special breathing gear to enter below-grade pits. As a result, machine manufacturers offer models where most or all of the machine is above grade. For very large machines, this means that the carrousel may be above the reach of the operator. These machines usually have automatic load and unload features.

4.2.7 Plug-Assist Means

Plugs used in heavy-gauge forming tend to require less temperature control than those in thin-gauge thermoforming. In many instances, the presses must be sufficiently robust to handle a few very large, relatively heavy plugs. Most heavy-gauge plugs are made of wood or syntactic foam and so plug temperature control is not required. Because these plugs may be many feet or meters above the shop floor, a means for easy adjustment and removal is required.

4.2.8 Pre-Stretching Means

Pre-inflation or pre-stretching of the heated sheet is quite common in heavy-gauge thermoforming. Two methods of pre-stretching are used. In the bubble stretching technique, the hot sheet is temporarily clamped over a lower mold element that acts as a blow box. The sheet is then inflated with the extent of inflation controlled with a photoelectric eye. The male mold is either lowered into the sheet from above or raised into the sheet from below.

The second method uses a vacuum or draw box. The sheet is drawn down with the extent of draw controlled with a photoelectric eye or a mechanical microswitch. The male mold is either lowered into the sheet from above or the female mold is raised around the sheet from below. Differential pressure across the sheet during pre-stretching typically ranges from 2 lb/in^2 (14 kPa) to 10 lb/in^2 (68 kPa). Both techniques require control of the stretched sheet shape as well as bubble deflation when the mold enters the inflated sheet.

4.2.9 Vacuum and Pressure

The vacuum system in heavy-gauge thermoformers may be either individual for a given machine (Fig. 4.7) or gauged for many machines. Vacuum surge tanks are required, particularly for deep cavity molds. The surge tank volume should be 6 to 20 times the combined volumes of the deepest cavity molds. Vacuum pumps should be capable of maintaining 28.5 in (725 mm or 35 torr) of mercury at their inlets. Surge tanks should be capable of maintaining at least 25 in (635 mm or 125 torr) of mercury at their inlets. Vacuum lines should be smooth with very few elbows and tees, and at least 4 in (100 mm) in diameter. Flexible lines are acceptable as long as they are smooth inside. The vacuum control valve should be a solenoid-actuated rotary ball valve. Ancillary blowers with inlets directly attached to the molds are used to help evacuate very deep cavity molds.

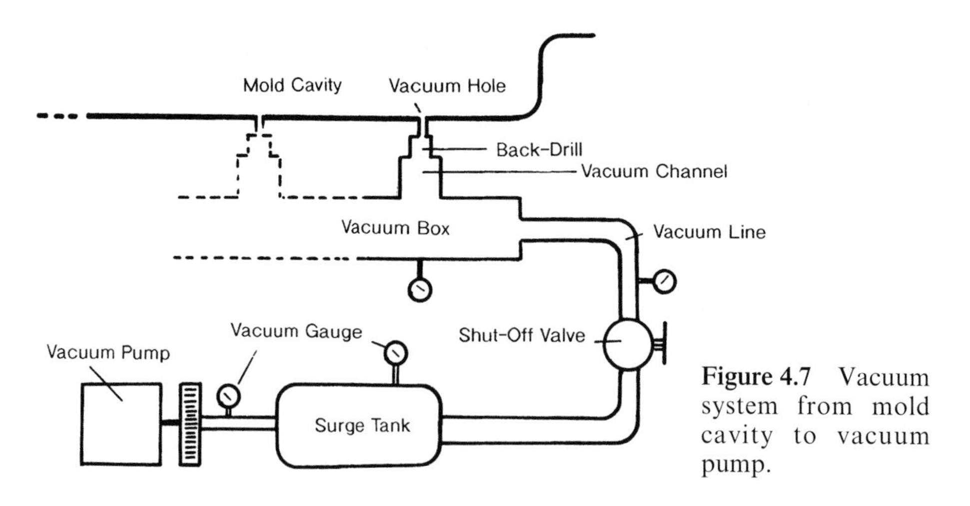

Figure 4.7 Vacuum system from mold cavity to vacuum pump.

Pressurized air is used extensively in thermoforming, even in all-electric machines. Care must be taken in using traditional compressed air when molding parts with critical surface requirements. Instrument air is recommended in general and micron-filtered air is required when the product is used in certain medical and food applications. Pressure requirements continue to increase. Pressure forming needs air pressures of at least 100 lb/in^2 (0.7 MPa), and the forming of multilayer, filled, or reinforced sheet requires air pressures of 200 lb/in^2 (1.4 MPa) or more. For very high pressures, piston-type compressors or boosters are added between the compressor and the individual machine pressure box.

4.2.10 Process Monitoring and Control

Despite the fact that the process allows extended time for measurement and control, heavy-gauge thermoforming has lagged thin-gauge thermoforming in the development of accurate process monitoring. Single-point, infrared monitors are currently used to measure sheet temperature. In some instances, the sheet is indexed from the oven based on its temperature, rather than the amount of time spent in the oven. In this case, inexpensive, long-wavelength, infrared monitors are used to measure the temperature at multiple points on the sheet's surface as it passes from the oven to the mold. Some machinery manufacturers also offer infrared thermographic devices that measure the entire sheet as it is heated.

Mold temperature control is common, but actual measurement of the mold's temperature during machine operation is not. Recently, machinery manufacturers have started providing individual watt meters mounted on heater banks to monitor electric power usage. Because the loss of a sheet in the oven is a major catastrophe, automatic protocols must be available to shut down the oven in case of fire, power overload, computer failure, brownout, power outage, light curtain interruption, and safety cage security breach. Additional important features are computers for data-logging, initial press set-up logging, temperature and power monitoring, and machine interruption and maintenance problem recording.

4.3 Thin-Gauge Form, Fill and Seal Machines

Many disposable, rigid, medical and food packages and certain point-of-purchase packages are produced in form, fill, and seal (FFS) equipment

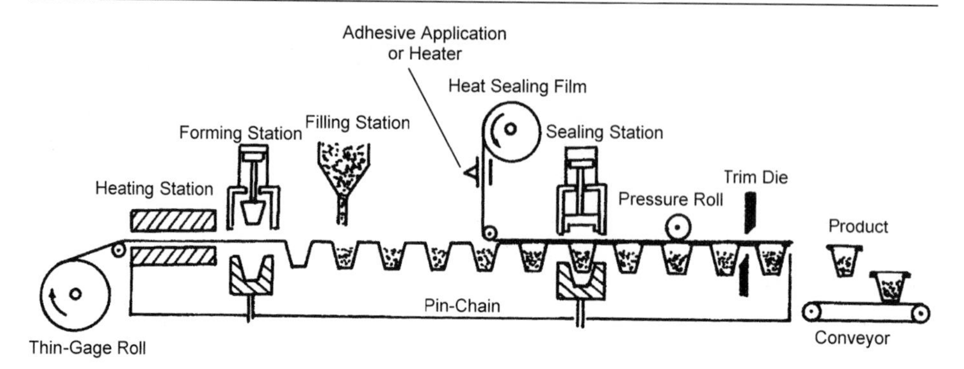

Figure 4.8 Form, fill, and seal (FFS) line.

(Fig. 4.8). Thin-gauge sheet, typically in the 10 to 20 mil (0.010 to 0.020 in or 0.25 to 0.5 mm) thickness range, is continuously fed from a roll to the heating station. For wide or heavier sheet, a pin-chain rail is used for conveyance. Narrow or thin sheet is just pulled through the machine from the web end.

Non-contact infrared heaters are used for heavier sheet, while direct contact heating is used for thin sheet. Contact heating can be accomplished in two general ways. For continuous contact heating, the sheet is wrapped over one or two heated rollers. For discontinuous contact heating, the sheet is brought in contact with the heater, either by sucking it against an electrically heated plate containing vacuum holes or by mechanically pressing the heater against the sheet. Because the sheet is typically not heated to traditional thermoforming temperatures, mechanical pressing is used, rather than pressure or vacuum forming.

The formed containers, still connected to the web, are then fed beneath a filling station. Filling is done manually in simple cases. In most operations, automatic dosing equipment is used. Depending on the nature of the product, this equipment can be very complex and expensive. After filling, the containers proceed to a sealing station. Sealing may mean simply snapping a lid on the container, but can involve gluing, heat sealing, or even ultrasonic sealing. The sealing material may be paper, plastic film, or even cardboard. Again, sealing is done manually in simple cases. In complex operations, heat and pressure is applied for the lid to adhere to the container.

The containers are then trimmed from the web by a reciprocating steel rule die punch. The containers may be stacked but are usually collected and placed in cartons. The web, along with excess sealing film, is reground and reprocessed, usually into products that are not used in medical or food contact applications.

Many FFS machines are built for specific applications. In certain industries, the machines must be stainless steel and may need to be steam-sterilizable. Although these machines incorporate thermoforming for manufacturing the rigid container, the major engineering effort in their design focuses on controlling the filling step and sealing the container.

4.4 Extrusion/Forming Lines

For dedicated thermoformed products, the extrusion process is often coupled with the thermoforming process. Refrigerator door liners are often produced on heavy-gauge extrusion/forming lines such as that shown in Fig. 4.9. Most extrusion/forming lines are found in thin-gauge forming operations, however, for the production of products such as drink cups and deli containers. Care must be taken to match the output of the extruder to the throughput of the thermoformer. The thermoforming process cycle should always dictate the overall process cycle.

Although the extrusion/forming line concept works well when everything is operating properly, significant problems can occur when coupling these two processes. For example, the extrusion process typically takes much longer to reach steady state than does the thermoforming process. As a result, at start-up and when process conditions change, the thermoformer is either idle or producing unacceptable parts.

If there is an interruption in the thermoforming cycle, the extruder must either be stopped or the sheet being produced spooled or reground. If the extruder is stopped for extensive periods, the polymer residing in the extruder barrel may be thermally damaged. The most common problem however, is mismatched capacities, with the output from the extruder being either too small or too great for the amount of product being formed.

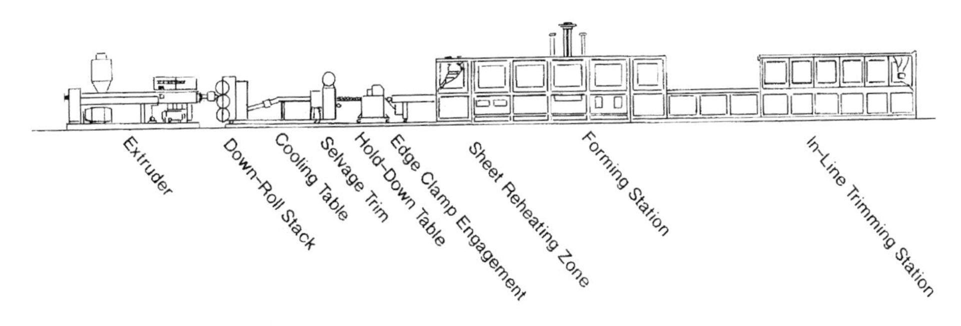

Figure 4.9 Extrusion/forming line, heavy-gauge.

Another problem may occur when both processes are under a single manager. The quality of the thermoformed part depends on the quality of the incoming sheet. If the sheet is fed directly from the extruder to the thermoformer, there is very little time to monitor sheet quality and reject sheet that does not meet requirements. Extrusion is also discussed in Chapter 10.

4.5 Matched Mold Forming Machines

Machines designed for matched mold forming fall in two categories, depending on the type of sheet to be formed. For low density foams, the primary concern is thermal conditioning the sheet prior to forming. Excessive heating results in cell collapse and poor product quality. Inadequate heating results in low secondary expansion and products with split corners and edges. Matched molds are used because these sheets cannot be heated enough to be vacuum-formed. Pressure forming cannot be used because excessive external pressure collapses the cells. To deal with these requirements, machines for low density foam molding are roll-fed with pin-chain rails that carry the sheet through the process. They have very long ovens with heaters that operate at relatively low temperatures. Because sheet sag is negligible at typical forming temperatures, oven heaters can be placed relatively close to the sheet surfaces. However, because metal rod heaters are used extensively in this application, thermal striping may necessitate that the heaters be placed quite far from the sheet surfaces. Press clamping forces are relatively low, with pressures commonly about 50 lb/in^2 (0.34 MPa).

On the other hand, when reinforced or highly filled polymers are formed, clamping pressures can exceed 200 lb/in^2 (1.4 MPa). As a result, special purpose machines for use with these polymers frequently utilize hydraulic pressures to hold the molds closed. These machines are usually shuttle machines.

5 Methods of Heating Sheet

The first step in forming a plastic part involves heating sheet to the proper forming temperature. Heat transfer and the methods of heating dominate the thermoforming technology because improperly heated sheet typically produces parts of poor quality, and the cost associated with heating represents a major portion of the final cost of the formed part.

In this chapter, we first discuss the three methods of heat transfer and how each of these are employed in thermoforming. We then examine characteristics of commercial heaters used in thermoforming. Finally, we discuss how energy input, oven geometry, and sheet characteristics combine to affect the heating of a sheet to its forming temperature.

5.1 Concepts in Heat Transfer

There are three general methods of transferring thermal energy between hot surfaces, or sources, and cold surfaces, or sinks: conduction, convection, and radiation.

5.1.1 Conduction

Heat transfer via conduction is energy transfer by contact between solids. Contact heating is used when polymer sheets are very thin. Conduction is also the primary way energy moves through plastic sheet and metal molds. Density, specific heat (heat capacity or enthalpy), and thermal conductivity are three thermal properties important in conduction. Thermal diffusivity, i.e., the ratio of thermal conductivity to the product of density and specific heat, is important in time-dependent heat conduction. Table 5.1 compares the thermal conductivities and thermal diffusivities of several polymers with similar properties of thermoforming mold materials. As is apparent, polymers are thermal insulators compared with metals. Conduction heat transfer into the polymer sheet from its surface is a controlling factor in the heating of thicker plastic sheets.

Table 5.1 Comparison of Thermal Conductivity and Thermal Diffusivity for Several Polymers and Mold Materials

Material	Thermal Conductivity		Thermal Diffusivity		Thermal Conductivity Relative to PS
	[Btu/ ft.hr.°F]	[$\times 10^{-3}$ kW/m.°C]	[$\times 10^{-4}$ ft^2/hr]	[$\times 10^{-4}$ cm^2/s]	
Polystyrene	0.105	0.180	29.7	7.66	1
ABS	0.07	0.12	25	6.45	0.67
Polycarbonate	0.121	0.207	33.0	8.51	1.15
Rigid PVC	0.100	0.171	32.5	8.39	0.95
LDPE	0.23	0.39	46	11.9	2.2
HDPE	0.29	0.50	55	14.2	2.75
PP homo-polymer	0.11	0.19	25	6.45	0.67
PET	0.138	0.236	36.8	9.49	1.3
Low density PS foam	0.016	0.027	80	20.6	0.15
Aluminum	72.5	124	18,850	4860	690
Steel	21.3	36.4	3,930	1010	200
Maple	0.073	0.125	104	26.8	0.7
Plaster	0.174	0.298	120	31.0	1.66
Syntactic foam	0.07	0.12	40	10.3	0.67

5.1.2 Convection

Convection heat transfer is heat transfer by contact between a fluid and a solid. Throughout thermoforming, the sheet is surrounded by ambient air. Energy is transferred when the air temperature differs from the sheet temperature. Energy transfer depends on air movement and the temperature difference between the fluid and the solid plastic sheet. As expected, energy transfer is low in quiescent or still air and is relatively high when the air is actively moved across the plastic surface. The proportionality between the temperature difference between the fluid and the plastic sheet and the amount of heat transferred is called the heat transfer coefficient. As seen in Table 5.2, the heat transfer coefficient value and the amount of heat transferred per unit time increase with fluid activity increase.

Table 5.2 Relative Comparison of Convection Heating Characteristics

Quiescent air	1
Air moved with fans	3
Air moved with blowers	10
Air and water mist	50
Water fog	50
Water spray	100
Oil in pipes	200
Water in pipes	500
Steam in pipes	3000

5.1.3 Radiation

Radiation heat transfer is an interchange of electromagnetic energy between hot and cold solid surfaces. Radiation pervades nature. Electromagnetic energy is usually characterized by the wavelength of the energy. The near infrared wavelength range is 0.7 μm to about 2.5 μm. Far infrared wavelength range is from about 2.5 μm to 100 μm. For most thermoforming processes, most of the energy is interchanged in the infrared range of 2 μm to about 15 μm. Figure 5.1 shows that at a given heater temperature, radiant energy output covers a wide range of wavelength, but peaks at a given wavelength. Table 5.3 tabulates temperatures associated with peak infrared wavelengths.

The efficiency of radiant heat transfer depends on the relative absorbing and emitting abilities of the heating source and the cooler plastic sheet. Absorptivity and emissivity are terms used to describe this efficiency. In most thermoforming operations, the dominant energy interchange is between the heater surface and the plastic sheet surface. In some cases,

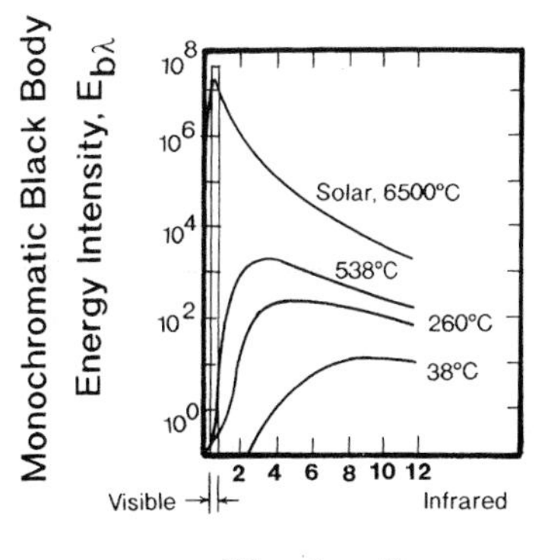

Figure 5.1 Wavelength-dependent radiant energy for heaters at various temperatures.

Table 5.3 Heater Temperature and Peak Infrared Wavelength

Temperature [°F]	Temperature [°C]	Wavelength [μm]
2000	1093	2.11
1500	816	2.66
1200	649	3.14
1000	538	3.57
800	427	4.14
600	316	4.92
400	204	6.06

energy is transmitted into or through the polymer sheet. In addition, even though the sheet is sandwiched between the heaters, neither the heaters nor the sheet are infinite in extent. As a result, the efficiency of energy interchange depends on geometric factors such as relative sheet-to-heater spacing and the extent of mechanical elements such as clamp frames, rails, and oven side-walls. Radiant heating is the fastest and most versatile means of heating sheet in thermoforming.

5.2 Common Heating Sources in Thermoforming

Table 5.4 lists common heating sources used in thermoforming. The simpler sources are used in prototyping and laboratory forming.

Table 5.4 Common Heating Sources

Hot air (including convection toaster ovens)
Hot water/steam
Sun lamps (drugstore variety)
Nichrome spiral wire (toaster wire)
Steel rod heaters
Steel or nichrome tape
Tungsten/halogen tube heaters
Quartz tube heaters (nichrome or tungsten wire or tape)
Steel plates with imbedded resistance wire
Ceramic plates with embedded resistance wire
Ceramic bricks with embedded resistance wire
Steel plates that re-radiate combustion energy from gas flame
Steel wire grids that re-radiate combustion energy from gas flame
Indirect gas combustion on catalytic beds
Direct gas combustion energy

5.3 Convection or Hot Fluid Heating

Recirculating hot air ovens are used when heating times are not critical or when sheet stock is very thick. Two types of ovens are employed. Horizontal ovens, similar to pizza ovens, are used when the sheet is not prone to excessive sag. The sheet is clamped in a frame or is supported on standoffs while in the oven. This allows for adequate air circulation around the sheet. The sheet may be moved in and out of the oven on rollers or rails or may simply be slid in and out manually. Usually, in horizontal ovens, the air is heated by blowing it across heating coils mounted in the tops of the ovens. Although there is some radiation heat exchange between the heating coils and the top surface of the sheet, the majority of the energy transfer is between the heated air and both surfaces of the sheet.

Vertical ovens are used when the sheet material tends to sag or where many sheets are heated simultaneously. Sheets, without auxiliary clamping frames, are hung from rails or tracks using either continuous clamps or individual clips. Sufficient tension and gripping area are required to ensure that the heated sheet does not extrude from the clamping device. In vertical ovens, the heating coils are placed along the oven sides and are baffled from the hanging sheets. Air is fan-blown across the heating coils and through the oven chamber. Many convection ovens use electrical resistance wire as the heating source. Indirect gas-fired heaters are also popular.

Direct gas-fired heaters, similar to those used in rotational molding ovens, are used if the polymer is not easily attacked by combustion products. Usually the oven air temperature is no more than 50°F (30°C) higher than the desired final sheet temperature. As with all convection ovens, the internal air temperature drops substantially while the oven door is open. Recovery to the set temperature usually takes several minutes. Simple proportional controls are used with electrically heated recirculating hot air ovens and on-off controls are used with gas heated ovens. Recirculating hot air ovens are used in aircraft windscreen, spa, and barrel and dome skylight production.

Water and oil are better heat transfer media than air, but recovery of the fluid and cleaning or drying the part after forming, limits their use as heating media in commercial forming. Boiling hot water can be used to soften such polymers as flexible PVC, semicrystalline ethylene vinyl acetate (EVA), and certain cellulosics. Water and oil are used as the hydraulic fluids in diaphragm forming and they are used extensively in cooling where mold temperature control is critical.

Steam is not normally used as a heat transfer medium. However, polystyrene has an exceptionally high water vapor transmission rate. As a result, steam permeates low density polystyrene foam sheet very readily, heating it to its forming temperature in seconds, without the risk of catastrophic cell collapse that can occur with infrared heating. Microcellular PS foam is heated and formed very successfully with steam.

5.4 Electric Heating

There are many types of commercially available electrically heated surfaces, from the simple nichrome "toaster" wire to tungsten filaments in quartz tubes, commonly called halogen heating elements. Typically, electric heaters are available in two general shapes. Round heaters, such as metal rod or calrod heaters and quartz tube heaters, emit thermal energy in all directions. Reflectors are needed to ensure the energy is directed in a parallel fashion toward the thermal sink, i.e., the plastic sheet. Flat heaters, such as metal plates and ceramic heaters, are already configured to direct the energy toward the plastic sheet. Some of the more common heaters are described below.

5.4.1 Round Heaters

At one time, nearly all thermoforming machines were equipped with rod heaters. The heating element is either a solid wire, a coiled wire, or a flat tape. The element is centered in a carbon steel or stainless steel sheath, with a compacted powdered inorganic oxide, such as magnesium oxide, filling the space between the element and the sheath. Electrical energy is converted to heat in the element, and that heat is then conducted through the oxide to the sheath. The sheath emits the energy in the form of infrared radiation. Rod heater output is available to 40 W/in^2 (60 kW/m^2).

Rod heaters are known for their ruggedness, longevity, and wide temperature range to 1500 °F (815 °C) or more. They are also known for their long heat-up times of tens of minutes, poor temperature control, and relatively rapid loss in efficiency. Although short rod heaters of 12 in (0.3 m) in length are available, typical lengths are 60 in (1.5 m) or more. Hairpin heaters are used when electrical connections are to be made from one side of the oven. To achieve uniform heating and to avoid thermal striping or local overheating, reflectors must be used with all rod heaters. Even with reflectors, rod heaters are usually placed many inches from the plastic surface. As a result, overall heating efficiency tends to be quite low. Proportional control is common.

Quartz heaters consist of metal tapes or wires centered in quartz tubes. Quartz glass is essentially transparent to infrared radiation. As a result, when the wire is energized, the wire heats and emits infrared radiation through the quartz tube. Reflectors are required with quartz heaters to ensure that the radiant energy is directed toward the plastic sheet. The common quartz heater uses a nichrome wire and the tube is either evacuated or filled with nitrogen. This heater has an output of around 40 W/in^2 (60 kW/m^2) and a temperature range to 1800 °F (980 °C).

Recently, halogen heaters have been developed for thermoforming. Tungsten wire is used and the tube is filled with a proprietary halogen gas. This heater has an output in excess of 50 W/in^2 (80 kW/m^2) and an upper temperature of 2000 °F (1090 °C) or more. Gold-plated reflectors are recommended for the higher temperature halogen heaters.

Quartz heaters are sought for their short heat-up times of a minute or less. They are quite fragile and the quartz surface can be etched by polymer off-gases, resulting in severe deterioration in heating efficiency. Temperature control is frequently on-off although predictive temperature control seems to work well.

5.4.2 Flat Heaters

The simplest and least expensive flat plate heater is a metal plate backed with serpentined nichrome wire imbedded in magnesium oxide. This is basically the flat plate version of the calrod heater. This type of heater is restricted to relatively low temperatures of up to 1000 °F (540 °C) and relatively low energy outputs of no more than 20W/in^2 (30 kW/m^2). Because this type of heater has many industrial uses such as in space heaters and hot plates, it is readily available and relatively inexpensive. It is frequently used as the energy source for home-built laboratory thermoformers. Temperature control is usually by rheostat or other proportional control. Heating times are usually on the order of tens of minutes.

Ceramic heaters are basically fire bricks or tiles with imbedded resistance wires. The heaters are made by partially filling a mold with ceramic slip, allowing the slip to dry to a green state, and then laying a heavy serpentine wire or coil on the green ceramic. Additional ceramic slip is poured over the wire to fill the mold. When the second pour is dried to a green state, the ceramic is then kiln-fired to produce a vitrified brick. Usually a glaze is applied to the brick and it is refired to achieve a high-gloss, smooth, chemically resistant surface.

When electrical power is applied to the wire, it heats. The heat is conducted to the surfaces of the brick. The energy then radiates toward the plastic surface. Ceramic heaters may include imbedded thermocouples for individual temperature control. Typically, ceramic heaters are useful to about 1300 °F (700 °C), with maximum energy outputs of about 40 W/in^2 (60 kW/m^2). Heating times are typically in minutes.

Ceramic heaters are available with flat or planar faces and with parabolic or spherical arc faces. As long as the ceramic heaters abut, there seems to be little difference in the performances of heaters with differing surface geometries. The bricks produced tend to be small, with surface areas of no more than 20 in^2 (130 cm^2). As a result, many tiles are required for very large thermoforming machines. Usually, blocks of tiles are modularly ganged or connected together to minimize wiring and control capacity. When tiles are ganged, one or more of the heaters should contain imbedded thermocouples for temperature monitoring and control. Proportional temperature control is usually used. Ceramic heaters are desired when careful zoning or patterning is required. Although ceramics are relatively rugged, thermal shock breaks them. Another disadvantage is that it is difficult to determine when a heater is not functioning.

Commercial flat panel heaters are designed specifically for thermoforming. Typically, the heating element is a heavy nichrome resistance wire or coil that is partially embedded in ceramic. The rear of the ceramic is heavily insulated with fiberglass or mineral wool. The face of the wire-ceramic is offset some distance from a flat panel. A quartz panel heater has a quartz surface panel. A metal panel heater has an aluminum, steel, or stainless steel panel. A quartz cloth heater has a woven quartz cloth panel. Radiant energy from the resistance wire passes through the quartz plate and quartz cloth. On the other hand, a metal plate re-radiates the energy absorbed from the resistance wire. The quartz plate and metal plate heaters are easy to clean, and all are rugged and relatively inexpensive.

Although panel heaters as small as 36 in^2 (230 cm^2) in surface area are available, common sizes are 144 to 576 in^2 (930 to 3716 cm^2). This large size inhibits pattern or zone heating. Maximum energy outputs are likely in the 20 W/in^2 (30 kW/m^2) range, with maximum surface temperatures of 1000 °F (540 °C). Units with 40 W/in^2 (60 kW/m^2) output and surface temperatures of 1700 °F (930 °C) are commercially available. Panel heaters tend to have long heating times, on the order of tens of minutes. Proportional temperature control is usually used, with external surface thermocouples providing temperature read-out.

5.5 Combustion Heating

Combustion is a well-explored means of generating heat. Natural gas and propane are the gases of choice for thermoformers. There are three types of gas heaters commercially available. The oldest is direct gas heating, where radiant energy from an open flame impinges directly on the sheet. Burners, similar to domestic gas furnace ribbon burners, are used. Because the energy output from direct gas heaters is high, perhaps in excess of $500\,W/in^2$ ($800\,kW/m^2$), only a few burners are used. To ensure some measure of heating uniformity, the burners are usually positioned very far, perhaps 30 in (0.75 m) or more, from the sheet surface. The combustion temperature of natural gas is 2300 °F (1260 °C). As with the domestic gas furnace burner, the only control is by turning the gas flow on and off.

Recently, indirect or catalytic gas heaters have been developed specifically for thermoforming (Fig. 5.2). A combustible gas-air mixture is introduced to the heater directly beneath a bed containing a special catalyst similar to the automotive exhaust catalyst or the catalyst used in camper heaters. Combustion occurs on and within the catalyst bed.

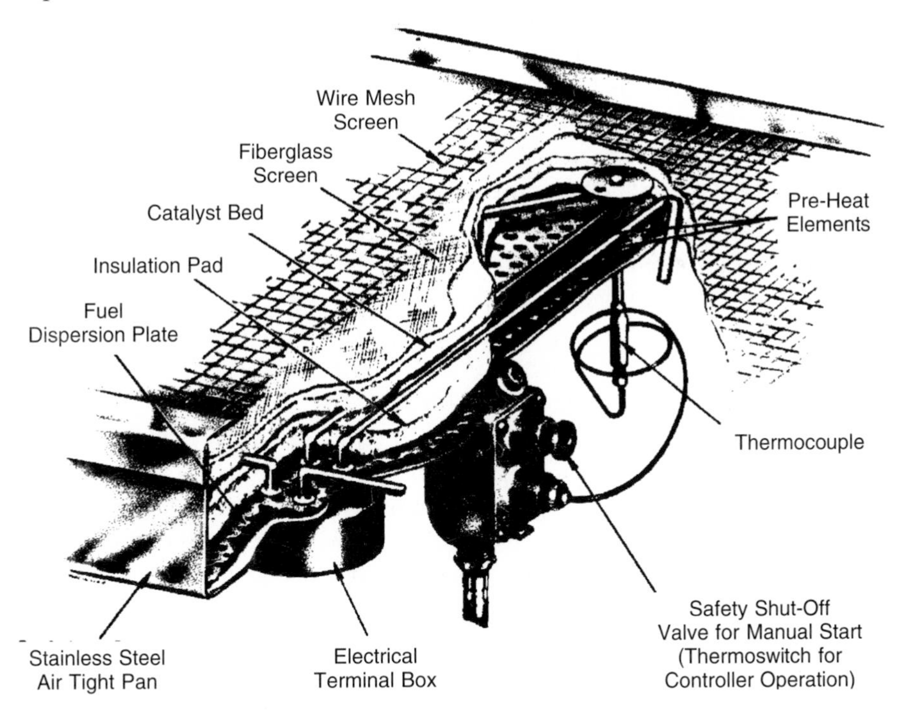

Figure 5.2 Catalytic gas heater (Valcan).

Catalytic heaters are sought for their low operating cost and very uniform surface temperature. They are restricted in size to about 576 in^2 (3716 cm^2) or larger. Catalyst deterioration has restricted surface temperatures to 800 °F (425 °C) and energy outputs are currently about 10 W/in^2 (15 kW/m^2). However, newer, more expensive catalysts promise higher surface temperatures. Until recently, the only temperature control was "on-off," as with direct gas burners. A gas flow control modulator now offers some degree of temperature turn-down. There is a very high initial installation cost involving not only many gas lines and controls, but also electrical energy for heating the catalyst bed to the proper combustion temperature. The typical heat-up time can be up to half an hour.

A third form of combustion heating has yet to be used extensively in thermoforming. The combustion products from ribbon burners impinge against metal screens or mesh or perforated ceramic plates. Re-radiating surfaces, sometimes called ported surfaces, provide very uniform surface temperatures. Temperatures to 2000 °F (1090 °C) are possible with surface energy outputs in excess of 200 W/in^2 (300 kW/m^2). Temperature control is achieved by turning the gas flow on and off. Because the wires or ceramic plates remain hot while the ported surface burner is off, radiant heat transfer continues, unlike in direct gas burners. Most commercial ported burners require gas pressures of about 5 lb/in^2 (30 kPa), in contrast to the 5 to 10 oz (2 to 4 kPa) gas pressures of direct gas burners and catalytic heaters.

All gas burners are designed for complete combustion, i.e., the generation of only water vapor and carbon dioxide as final products. Improper adjustment may lead to the generation of small amounts of soot and carbon monoxide, as well as other combustion gases. As a precaution,

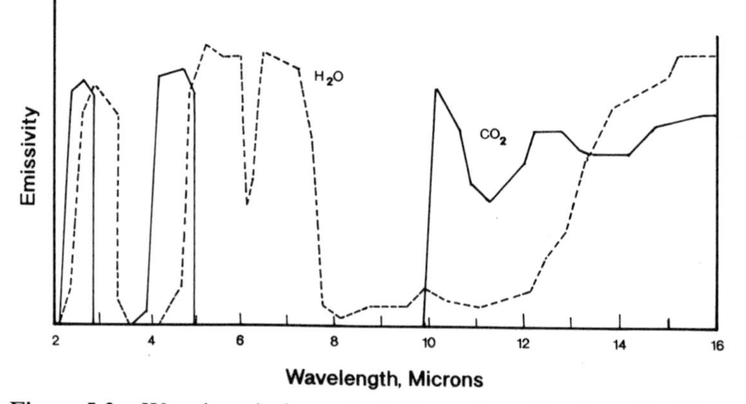

Figure 5.3 Wavelength-dependent absorption of carbon dioxide and water-vapour.

combustion ovens should be properly ventilated. Water vapor and carbon dioxide are known absorbers of infrared energy (Fig. 5.3). If the concentrations of these gases are very high, energy is preferentially absorbed by the gases, then re-radiated at the lower temperature of the gas. As a result, the overall heat transfer effectiveness is reduced. For catalytic heaters and for most open ovens used in thermoforming, the concentrations of water vapor and carbon dioxide rarely get high enough to cause major reductions in heat transfer effectiveness.

5.6 Selection of the "Correct" Heater

There is no "correct" heater. Many factors enter into the decision to purchase heaters. Certainly sheet geometry (width, length, and thickness), is important, as is the anticipated variation of sheet dimensions over the life of the thermoforming machine or heater. The type of forming envisioned is important, because roll-fed forming may include many heating stations ahead of the forming station, while a heavy-gauge forming machine may have only one heating station. Economic factors include the following:

- Day-to-day running cost. This is one of the major selling points for gas heaters. For the same unit of power delivered, natural gas costs in the US are typically 25% those of electricity.
- Maintenance cost. Rod heaters run years without substantial maintenance problems. Rod heater efficiencies deteriorate rapidly unless parabolic reflectors are replaced and rod surfaces are sanded and cleaned regularly. On the other hand, quartz tube heaters require periodic cleaning and careful treatment to minimize failure. Undetected burned-out ceramic heaters cause processing problems, so preventative maintenance is necessary. Catalytic gas heater surfaces must be carefully inspected for hot spots regularly and gas burners must be periodically checked for complete combustion. The fragility of heaters figures indirectly in maintenance costs.
- Initial installation cost. Large panel heaters require few power connections in contrast to small ceramic heaters which require substantial wiring. Gas heaters require both gas and electrical connections.
- Versatility of the heater. A large effective temperature range may be important if many types of polymers are to be formed. Rapid heat-up may be important if many mold changes are anticipated. Ease in zoning or patterning the heater output may be important if many sizes of sheet are to be heated or if parts are very complex. While these factors are

intrinsic, if a chosen heater configuration is not flexible, work may be lost.

Because all electric and gas heaters operate in the 300 to 1000 °F temperature range and all heater surfaces are in contact with oven air, all heaters generate both radiant energy and convective energy. Generally, if the heater configuration requires large open areas between individual heater elements, the heater bank must be placed some distance from the sheet surface to achieve uniform energy input to the sheet. As a result, air circulation between the heater bank and the sheet surface and between the individual heater elements is enhanced. Because hot air rises, the air heated by the upper heater bank remains near the heater while the air heated by the lower heater bank collects at the underside of the sheet. This is true whether the heating elements are electric or gas.

For most electric heaters, the energy output is essentially independent of wavelength. In other words, the emissivity is essentially constant at about 0.90 to 0.95. The emissivity is the fraction of output energy compared with a perfect emitting source called a black body that has an emissivity of 1.00. Gas heaters appear to have wavelength-dependent emissivities with average values

Table 5.5 Comparative Rating of Common Infrared Heaters Used in Thermoforming

Item	Metal Rod	Ceramic	Quartz Tube	Catalytic Gas
Radiant efficiency	55%	95%	60%	80%
Initial efficiency	Low	Medium	High	High
Maximum temperature				
	1400 °F	1400 °F	1600 °F	800 °F
	760 °C	760 °C	870 °C	425 °C
Longevity	High	Medium	Med/Low	Medium*
Manufacturing cost	Low	Medium	Medium	High
Installation cost	Low	Medium	Medium	High
Retrofit capability	Excellent	Good	Good	Poor
Operating cost	Low/Med	Medium	Medium	Lowest
Power response	Low	Medium	High	Low
Loss of effectiveness	High**	Medium	High	Med/High*
Pattern/zone capability	Poor	Excel/Good	Excellent	Poor
Chemical attack	High	Low	Medium	Low/Med
Breakage	Low	Low/Med	High	Low
Temperature control	Good/Poor	Good/Excel	Excellent	Poor

* Long-term data on catalyst longevity, loss in performance are unavailable
**Improves to Medium with programmed replacement of reflectors

Table 5.6 Disadvantages of Infrared Heaters Listed in Table 5.5

Metal Rod	Rust, difficulty detecting burnout, poor patterning, very slow power response, rapid loss in efficiency unless reflectors are replaced
Ceramic	Cost, best with PID controller, difficulty detecting burnout, tends toward brittleness
Quartz Tube	Very brittle, glass can be etched, requires careful cleaning, high initial cost
Indirect Gas	Excessive plumbing, poor temperature control, still requires electrical system for preheat, oven area may need venting, exhaust

of about 0.85 to 0.90. The small amount of wavelength dependency may be attributed to the presence of water vapor and carbon dioxide. Tables 5.5 and 5.6 give a comparison of common infrared heaters used in thermoforming.

5.7 A Comparison of Widely Used Heaters

Many rankings of heater types for various thermoforming operations have been published. One ranking is given as Table 5.7. It must be understood that any comparison must take economics and common sense into account.

5.8 Heating Cycle Times

There are many factors that influence the time required to heat a specific polymer to its forming temperature. These are reviewed here.

5.8.1 Forming Temperature

Generic guides for various polymers frequently list upper and lower forming temperatures. For amorphous polymers such as PS, the listed forming temperature range may be 100 °F (55 °C) or more. For crystalline polymers such as HDPE, the listed forming temperature range may be less. However, for specific applications, specific polymers form well over relatively narrow forming temperature ranges. Part wall uniformity, sheet sag, web formation, mold surface replication, vacuum hole dimples, mold and plug mark-offs, trim edge, three-dimensional corner brittleness, shrinkage, and residual stress in the finished part are all affected by the actual sheet temperature at the time of forming.

Table 5.7 Ranking of Heater Types For Various Thermoforming Operations (Heaters are numbered in order of preference)

Thermoforming Operation	Direct Contact	Hot Air	Ceramic	Metal Panel	Glass Panel	Quartz Cloth	Quartz Tube	Metal Rod	Gas Catalytic	Halogen
Form-Fill-Seal	1		2							
Thin-Gauge, Roll-Fed, One-Side Heating	1		2			3				
Thin-Gauge, Roll-Fed, Simple										
Top			3	4		1		2		
Bottom			4	2		3		4		
Thin-Gauge, Roll-Fed, Complex										
Top			1				2			
Bottom			1	2						
Heavy-Gauge, Sheet-Fed, Very Thick										
Top		1							2	
Bottom		1							2	
Heavy-Gauge, Sheet-Fed, Shallow										
Top					4	1	5		2	3
Bottom				1				2	3	
Heavy-Gauge, Sheet-Fed, Deep										
Top			1			4	2			3
Bottom			2	1					3	
Heavy-Gauge, Sheet-Fed, 4-Station										
1st Top				2				3	1	
1st Bottom				2				1	3	
2nd Top			1				2			3
2nd Bottom			2	1					3	

The upper forming temperature is usually dictated by polymer discoloration, excessive smoking, thermal degradation, or excessive sag. The lower forming temperature is usually determined by whether the polymer can be drawn into the shallowest of molds or by excessive residual stress in the final part. Some guides list orientation temperature values as being the temperature at which the polymer can be stretched 375%. Table 5.8 gives some typical normal forming temperatures for thermoformable polymers. Understand that actual forming temperatures may differ significantly from these values.

Table 5.8 Normal Forming Temperatures for Thermoformable Polymers

Polymer	Temperature	
	[°F]	[°C]
Polystyrene [GPPS]	300	149
ABS	330	166
Rigid PVC	280	138
PMMA [Acrylic]	350	177
Polycarbonate [PC]	375	191
HDPE	295	146
Polypropylene	310	154
APET	300	149
40% GR PP	400	204

5.8.2 The Nature of Energy Uptake

Energy is transferred from the heating sources to the surface of the sheet by a combination of convective air motion and infrared energy transfer from heaters. All the energy is absorbed in the first 10 to 30 mils (0.010 to 0.03 in or 0.25 to 0.75 mm) of the sheet surface. This energy is conducted into the core of the sheet. As a result, the temperature at the sheet center is always lower than that at the sheet surface. As the sheet thickness increases, the temperature difference between the sheet center and sheet surface increases, all other factors constant. As the sheet thickness increases, conduction into the center of the sheet becomes more and more important to ensure that the sheet surface does not exceed the upper forming temperature before the center is sufficiently hot.

In short, for thin sheet, energy transmission into the sheet is significantly less important than energy transmission to the sheet. For very thick sheet, the opposite is true. For thin sheet, short wavelength, high-temperature, infrared heaters such as quartz tube heaters minimize heating cycle time. For thick sheet, long wavelength, low-temperature heaters such as catalytic gas heaters minimize sheet surface overheating. For very thin sheet, care must be taken to maximize radiant absorption in the sheet. If too much radiant energy is transmitted completely through the sheet, it heats very inefficiently. Direct contact heating may heat the sheet more quickly. For very thick sheet, even low-temperature radiant heaters may provide too much surface energy. Forced air convection ovens may be the only alternative.

5.8.3 Polymer Characteristics

There are two major factors that dictate how a specific polymer heats to its forming temperature. The first is the total amount of energy needed to raise the polymer from room temperature to the forming temperature. As noted in Fig. 2.1, crystalline polymers such as HDPE require more energy than amorphous polymers such as PS. If the energy to the sheet is constant, it takes longer to heat crystalline polymers than amorphous polymers.

In addition, thermal diffusivity is the polymer thermal characteristic that most affects time-dependent or transient heat conduction from the sheet surface to its core. Polymers with higher values of thermal diffusivity have lower temperature differences from the sheet surface to the core than polymers with lower values, all other factors constant. As seen in Table 5.1, this value for ABS is slightly less than that for PS. Because the energy uptake required to heat ABS to its forming temperature is about the same as that for PS, the temperature difference through the ABS sheet should be slightly greater than that through the PS sheet, again if all other factors are constant and as long as conduction into the sheet controls.

On the other hand, the thermal diffusivity value for HDPE is nearly twice that for PS and so the temperature difference through the HDPE sheet should be about half that for PS. The heating rate for HDPE can be increased if the temperature difference for PS is acceptable for HDPE. Despite this advantage, the general effect is that HDPE heats to its forming temperature at a rate less than that for PS, because it takes twice as much energy to heat HDPE to its forming temperature as it takes for PS, again all other factors being constant.

5.8.4 Geometry

Sheet and heaters are characterized as having finite length, width, and sheet-to-heater spacing. In addition, they are flat or planar, at least initially for the sheet. The finiteness of the heaters and the sheet and the presence of assorted oven elements such as side walls and rails, affect the efficiency of energy transfer between the sheet and the heaters. The view factor concept is aptly named, because heaters radiate to everything in view. For energy to be transferred between the sheet and the heaters, they must "see" each other. For the maximum amount of energy to be transferred, they must see as much of each other as possible. Obviously, if the heaters are very close to the sheet and the sheet and heater lengths and widths are very large, they essentially see only each other. As a result, the view factor has a value of about one. On the other hand, if the sheet dimensions are very small compared with the oven dimensions and the distance between the sheet and heaters is very large, the heaters see not only the sheet but oven walls, rails, and other non-sheet surfaces. This means that the effective energy transmission is quite low and the value of the view factor is substantially less than one.

If all heaters are set to the same temperature and the sheet is approximately the same temperature, an energy dome results (Fig. 5.4). Because the heaters above and below the center of the sheet see more sheet than those at the sheet edges or corners, a greater amount of energy is transferred to the sheet. As a result, the sheet is hotter in the center than on the edges or corners. To compensate for this, center heater temperatures are reduced and corner and edge heater temperatures are increased. This is called zoned heating or zonal heating. When large surface area panels or rod heaters are used, welded stainless steel or nickel wire screen or mesh is cut and placed in areas where less energy is desired. This is called pattern heating.

One additional factor is important. When the sheet sags toward the lower heater, the overall view factor changes. Furthermore, the local view

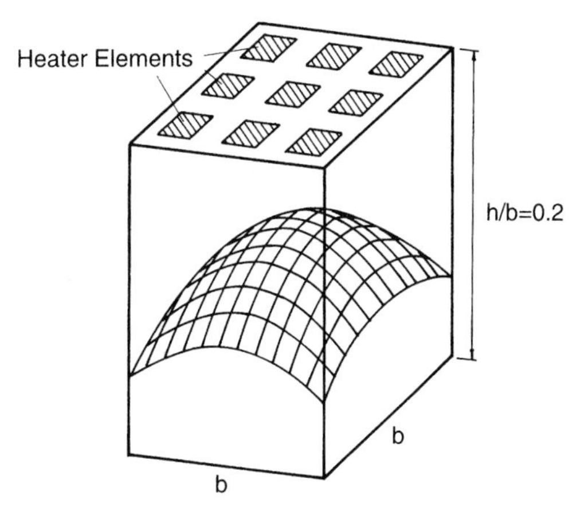

Figure 5.4 Energy dome resulting from uniform energy input to a finite dimensioned sheet.

factor, being the energy transfer between an individual heater and a corresponding region on the sheet surface, also changes. Locally, the lowest portion of the sagging sheet can very quickly increase in temperature. In short, the general heating pattern is altered when the sheet sags. Placing the lower heater far away from the bottom sheet surface helps mitigate this problem at the expense of low effective energy transmission.

5.8.5 One-Sided vs. Two-Sided Heating

Sandwich heating, with heaters over and below the sheet, is a commonly recommended practice. Conduction energy transfer is governed not only by the polymer thermal diffusivity but by the square of the sheet thickness. If the sheet is heated on only one side, it takes about four times longer for the sheet to reach a forming temperature (Fig. 5.5). For very thin sheet in form, fill, and seal operations, the heating time is not as critical as having the sheet at exactly the right temperature for press-forming. In turn, this temperature is not as important as the critical aspects of filling and sealing. As a result, many FFS formers heat the sheet only on one side.

In simultaneous twin-sheet forming, two sheets are clamped in a common frame, with air introduced between them to keep them separated. The frame is then introduced to the oven where the top heater heats the top surface of the top sheet and the bottom heater heats the bottom surface of the bottom sheet. The relative ease of simultaneously forming both sheets frequently

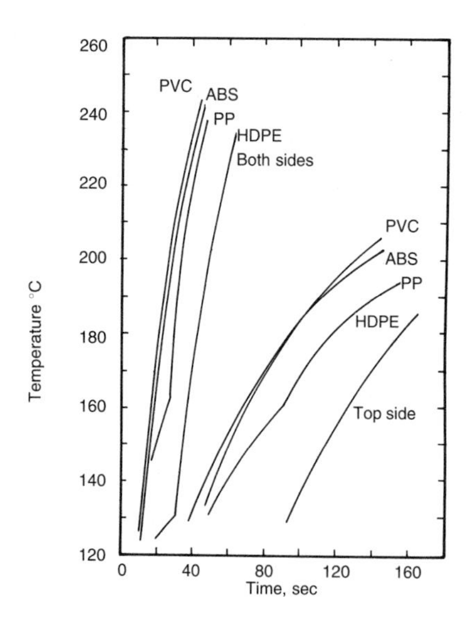

Figure 5.5 One- v. two-sided heating of various polymers 40 mil (0.040 in or 1 mm) sheet; top heater 450 °C at 20 kW/m^2; bottom heater 400 °C at 12 kW/m^2.

compensates for the extended heating time. Some thermoformers who are more concerned about surface appearance and part quality than cycle time mount the sheet directly on or over the mold, move the heater over the top surface of the sheet and heat from one side only. When the sheet is thoroughly heated, the oven is retracted, the sheet contacts the mold rim and vacuum is applied. Spas and golf-cart shrouds are two products made this way.

5.8.6 Heating Cycle Time Prediction

The best way to determine heating cycle time is to actually heat a sheet of plastic under controlled conditions to the desired forming temperature. Previous experience heating similar polymers in the same oven configuration is the next best way. Typically, for thin sheet of less than about 30 mils (0.030 in or 0.75 mm), the heating cycle time decreases about linearly with increasing heater temperature over a heater temperature range of about 100 °F (55 °C) or so. Typical thin-gauge heating cycles are on the order of seconds to tens of seconds. The heating time for 30 mil (0.030 in or 0.75 mm) PVC sheet is on the order of 6 to 8 seconds, with a heater temperature of about 1400 °F (760 °C). Because the controlling resistance to heating is the intensity of the radiant heat, very high heater temperatures can be used. However, the rate of heating is not limited by the output of the heaters, but by the mechanics of moving sheet from the oven to the forming press, closing the press, forming, opening the press, and repeating the cycle. The lower practical heating cycle is probably on the order of a few seconds.

For sheet greater than about 0.375 in (9.5 mm) in thickness, the heating rate is controlled by the sheet surface temperature. The best cycle time is achieved when the sheet exits the oven with its surface temperature below the upper forming temperature and its centerline temperature above the lower forming temperature recommended for the polymer. Typical heating times for 0.125 in (3 mm) styrenic sheet are in the order of 20 to 45 seconds with heater temperatures of about 700 °F (370 °C). Typical heating times for 0.375 in (9.5 mm) styrenic sheet are in the order of 3 to 6 minutes. Computer programs to aid in determining these conditions are given in the literature.

5.8.7 Equilibration

While it is not necessary or practical to have the sheet at a uniform temperature across its thickness when it exits the oven, it is recommended that the sheet temperature across its thickness be as uniform as possible when it is stretched against the mold surface. This is achieved almost immediately when thin-gauge sheet is removed from the oven. As with conduction-dominated heating, the time to achieve uniform temperature across the sheet increases in proportion to the square of sheet thickness. Some temperature equilibration occurs as the sheet is transferred from the oven to the forming station. In addition, to improve forming, heavy-gauge sheet is sometimes allowed to "rest" for several seconds to as much as a minute before it is formed.

6 Sheet Stretching and Cooling

Thermoformed parts are manufactured by heating a sheet of plastic until it is rubbery, stretching it until it contacts a mold surface, and then allowing it to cool to retain the shape of the mold. This chapter considers the stretching or elongational characteristics of the rubbery plastic and the extent of stretching needed to produce useful formed parts. The importance of cooling is also reviewed.

6.1 The Concepts of Modulus and Stiffness

The stiffness of a given formed part is the product of the shape or geometry of the part and the elastic property of polymers called modulus. For a given part geometry, stiffness is dependent on the polymer's elastic modulus at the temperature at which the polymer is used. As with most polymer properties, the elastic modulus of any given plastic is temperature-dependent. As seen in Fig. 6.1, the moduli for amorphous polymers, such as PMMA, PC, PVC and PS, decrease with increasing temperature. The modulus for a given polymer drops two or more decades as the polymer is heated through its glass transition temperature range. As the polymer is heated above its glass transition temperature range, the drop in modulus slows dramatically for tens of degrees' increase in temperature. This region is sometimes called the plateau region and it is the technical definition of the thermoforming window.

For semicrystalline polymers, such as polypropylene (PP) in Fig. 6.1, the first transition is the glass transition temperature. Although there is a measurable drop in modulus as the polymer is heated through this transition, the drop is usually less than a decade.. Even though the amorphous regions of the polymer become rubbery, the crystallites do not. They provide the polymer's structure with strength. As heating continues to the crystalline polymer's melting temperature, the drop in modulus becomes dramatic. The drop may be only three decades or so, as it is with

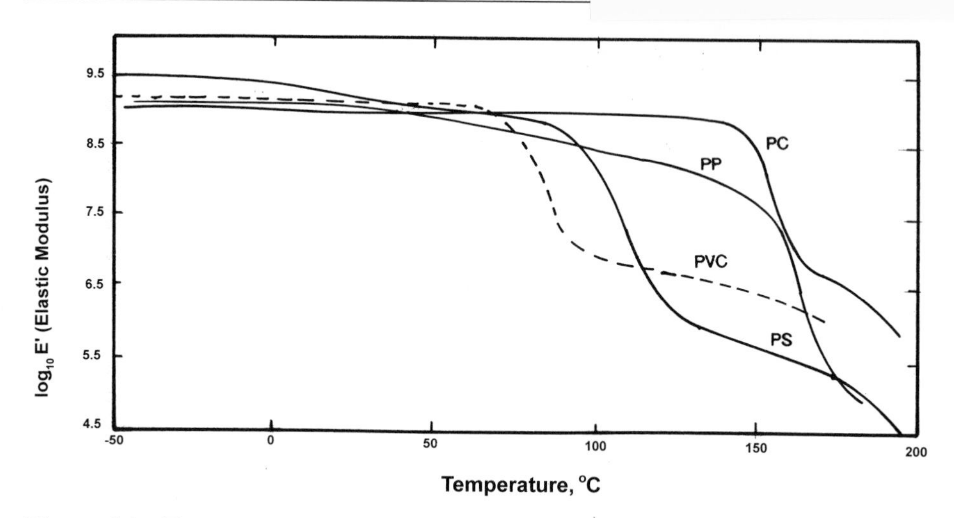

Figure 6.1 Temperature-dependent elastic modulus for several thermoformable polymers.

polyethylene, but it can be as much as five or six decades, as with the nylons (polyamides). If the elastic modulus value beyond the polymer's melt temperature is still relatively high, the polymer is said to be an elastic liquid. If the modulus of the molten polymer is low or if there is no plateau region, sheet may not hold together during heating. In short, such a polymer may not be thermoformable.

Elasticity is demonstrated when a stretching force on a polymer is released and, the polymer returns instantaneously to its original shape, much like a rubber band. To prevent this from happening in thermoforming, the stretched polymer is "frozen" into the desired shape by cooling it until it is glassy or crystalline. The original shape, i.e., a flat sheet, is recovered only when the formed part is reheated above the glass transition or melting temperature of the polymer. Fluidity is demonstrated when stretching force on polymer is released and there is no recovery. In other words, when the formed part is reheated above its glass transition or melting temperature, it remains in the shape it had prior to the release of the stretching force.

Figure 6.2 illustrates mechanically the relationship between the solid property of elasticity and the fluid property of viscosity. The spring represents the elastic modulus and the viscous dashpot or damper represents the fluid viscosity. Usually the elongational or extensional viscosity is used in thermoforming. The viscous effect is small below the amorphous polymer glass transition temperature and the crystalline

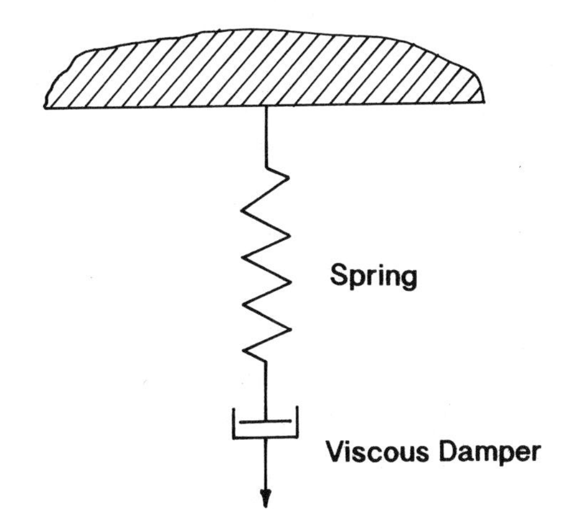

Figure 6.2 Series spring and viscous damper model (Maxwell viscoelastic model).

polymer melting temperature. The elastic effect is small at high temperatures, where the polymer is fluid-like. In between, as in the plateau region, the relative effects shift with temperature. Forming rates are quite high for most thermoforming processes. As a result, the plastic behaves as if it is elastic. In certain parts of the process, the stretching rates are slow enough that the viscous effects are seen. Plug assist stretching and sag are examples where the effects of viscosity are observed.

6.2 The Concepts of Stress and Strain

Differential pressure is used to stretch the heated plastic sheet. If the polymer is sufficiently supple, the differential pressure may be achieved by sealing the hot sheet against the mold surface and evacuating the air from between the sheet and mold. Higher pressures are used in pressure forming and filled and reinforced sheet forming. The differential pressure represents the stress applied to the sheet, with units of lb/in^2 or MPa. The extent of stretching is usually given in uniaxial-elongational percent, or stretching in one direction, and is technically called strain.

Stress-strain curves describe only the elastic behavior of a polymer. Figure 6.3 is a schematic of the temperature-dependent stress-strain curve, illustrating typical polymer elastic characteristics. Below the glass transition temperature range, there is little elongation before breaking. Above the glass transition temperature range, the polymer may show yielding before breaking. At higher temperatures, the stress-strain curve may be described simply as "power-law." At even higher temperatures, the elastic effect may

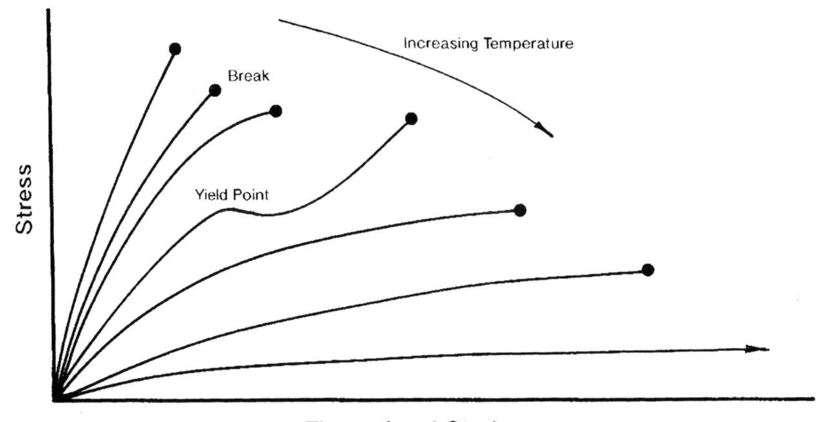

Figure 6.3 Temperature-dependent stress-strain curves.

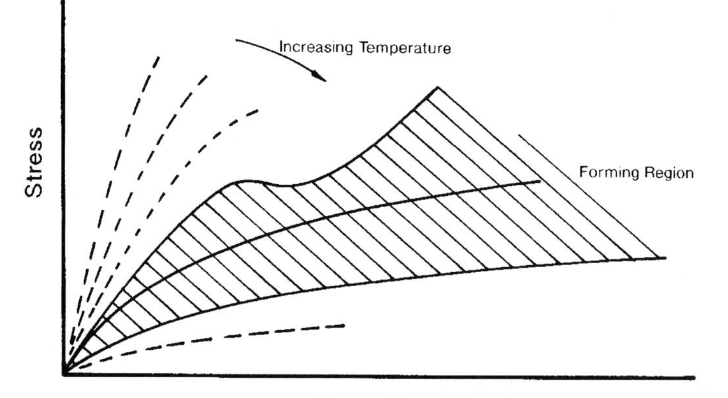

Figure 6.4 Temperature-dependent stress-strain curves with forming temperature overlay.

be so small as to be unmeasurable. In Fig. 6.4, the forming window, obtained either from tables or from experience, is overlaid on the schematic stress-strain curve.

6.2.1 Molding Area Diagram

The applied differential pressure represents the maximum stress that can be applied to the sheet during forming. The horizontal line in Fig. 6.5

represents this maximum stress. This line can represent absolute vacuum, expected vacuum, normal pressure forming, high-pressure forming, or even sag. As expected, the maximum amount of stretching increases with increasing temperature. When the stress-strain data for a given polymer are depicted in this fashion, the cross-hatched area below the horizontal line represents the thermoforming version of a molding area diagram.

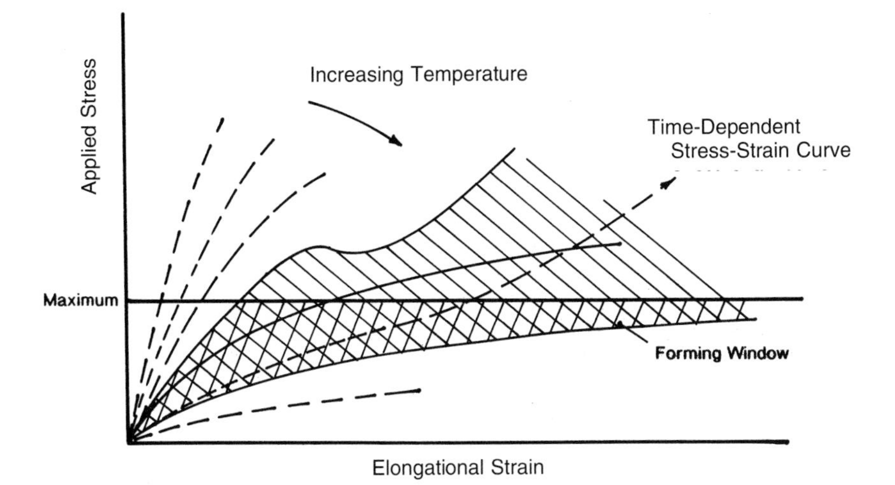

Figure 6.5 Temperature-dependent stress strain curves with forming temperature overlay and maximum applied stress (cross-hatched area is thermoforming molding area; dashed line represents effect of cooling during forming).

6.3 Pneumatic Prestretching

When the plastic sheet is hot, the pressures needed to stretch it are relatively low. Pressures on the order of 1 to 10 lb/in^2 (70 to 700 kPa) are used in pneumatic prestretching, whether in pre-blowing or vacuum box drawing. Table 6.1 gives representative inflation pressures for several thermoformable polymers.

6.4 How Sheet is Stretched Against a Mold Surface

As seen in Fig. 3.1, plastic sheet is stretched differentially. That is, when a portion of the deforming sheet touches a portion of the mold surface, it sticks and does not stretch anymore. The portion of the sheet that is free of

Table 6.1 Inflation Pressure Ranges for Thermoformable Polymers

Polymer	Inflation Pressure Range		Inflation Temperature Range	
	[lb/in^2]	[kPa]	[°F]	[°C]
PS	2–4	14–28	275–300	135–150
ABS	1.5–4	10–28	285–300	140–150
PMMA	7–10	48–70	320–355	160–180
Rigid PVC	1.5–3	10–21	240–285	110–140
Flexible PVC	1–3	7–21	240–285	110–140
PC	6–10	41–70	350–375	170–190
PET	2–4	14–28	275–320	135–160
LDPE	1–3	7–21	255–290	125–145
HDPE	1–3	7–21	265–300	130–150
PP	1–2	7–14	300–330	150–165

the mold continues to stretch until either the stretched sheet covers all the mold surface or the sheet becomes too stiff to allow further stretching. As a result of this differential stretching, the sheet is thickest where it first touches the mold and thinnest where it last touches the mold. Wall thickness variation for a relatively deep draw cup is shown as the solid line in Fig. 6.6, illustrating that the material at the rim is the thickest and the material in the corner is the thinnest.

Plugs are used to redistribute the polymer, as shown as the dashed line in Fig. 6.6. The sheet is drawn nearly uniformly between the bottom edge of the plug and the rim of the mold. The plug touches only that part of the sheet that becomes the bottom of the cup. The effect is to make the wall thickness of the cup more uniform and the corner thicker. The bottom of the cup is also thicker because the plug chills the sheet, making it more resistant to stretching.

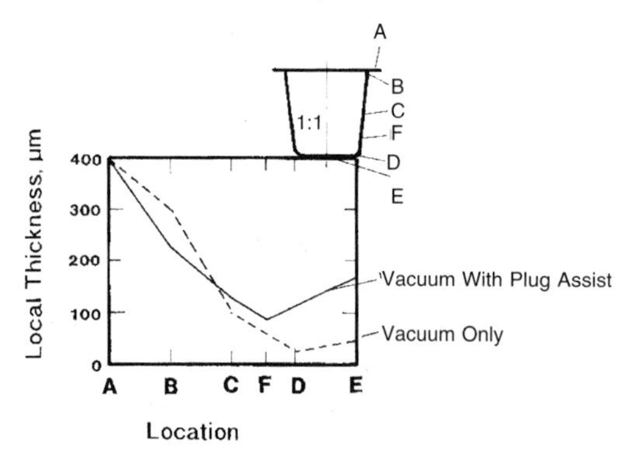

Figure 6.6 Wall thickness distribution with and without plug assist (16 mil, 0.016 in, or 0.4 mm MIPS).

6.4.1 Draw Ratios

There are three common definitions for overall draw ratio. The areal draw ratio is the ratio of the area of the formed part to that of the sheet used to form the part. The average reduced sheet thickness is the reciprocal of the areal draw ratio. The linear draw ratio is the ratio of the length of a line drawn on the formed part to that of the line drawn on the sheet before forming. The height-to-dimension ratio, written as H:D, H:d or H/d, is the ratio of the measured height of the formed part to the greatest dimension across the opening of the formed part. H:D is commonly used in Europe and is usually reserved for simple symmetric parts such as cups (see Fig. 6.6 where the H:D for the cup is 1:1). Figure 6.7 illustrates these draw ratios for the symmetric can-type container shown in Fig. 6.8. Maximum areal draw ratios for several formable polymers are given in Table 6.2.

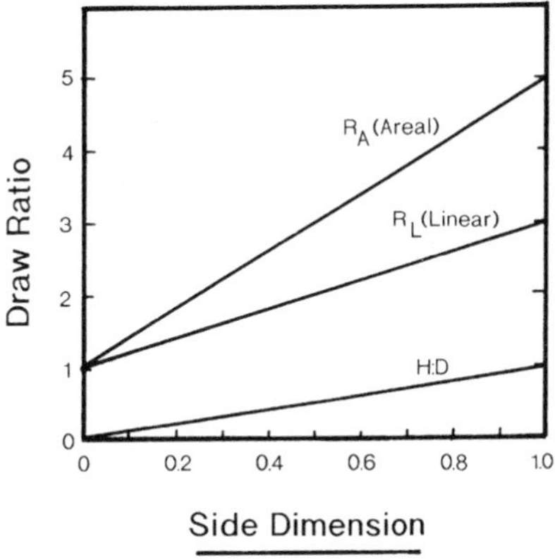

Figure 6.7 Draw ratios for cylindrical female part, I.

Table 6.2 Maximum Areal Draw Ratios for Thermoformable Polymers

Polymer	Maximum Areal Draw Ratio	Temperature [°F]	Temperature [°C]
PS	8.0	250	123
ABS	5.5	330	165
PMMA	3.4	310	155
Rigid PVC	4.3	255	125
Flexible PVC	4.2	230	110
LDPE	6.0	285	140
HDPE	6.5	330	165
PP	7.5	350	175

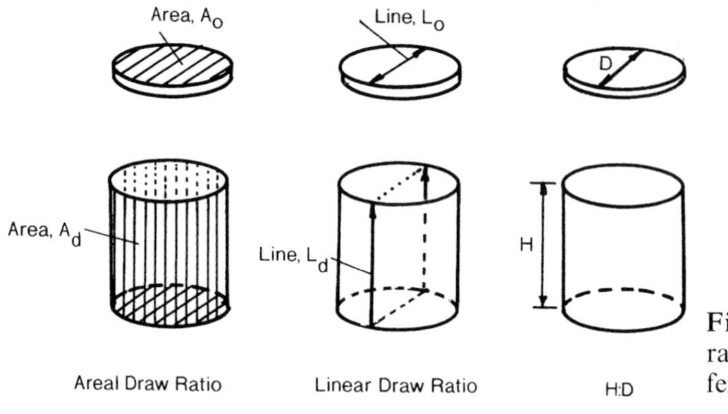

Figure 6.8 Draw ratios for cylindrical female part, II.

6.4.2 The Utility of Draw Ratios

Overall draw ratios have no value in thermoforming. From a practical viewpoint, some of the most difficult-to-form parts are shallow draw-ratio parts. Parts that have picnic-plate configurations or very large radii-of-curvature are examples. The draw ratio value gives no indication of wall thickness variation across the part. An overall area draw ratio value for a 60° cone is exactly 2, even though the sheet thickness is essentially zero at the tip of the cone. Overall draw ratios relate only to the mold geometry. The draw ratios for plug assisted parts and unassisted parts are identical, as are the draw ratios for different polymers at different temperatures. The draw ratio cannot give important information such as the temperature-dependent, smallest feasible corner radii for various polymers as given in Fig. 6.9.

6.4.3 Input for Computer-Aided Design

Geometric guidelines and mathematical design models, such as finite-element analysis, are discussed in Chapter 9. Because mathematical design models require a relationship between the applied stress and the resulting strain, the stress-strain relationships of Fig. 6.4 must be quantified. Two mathematical models have been used. The oldest, called the Mooney-Rivlin model, was developed for natural rubber elasticity and is used extensively in such fields as tire design. The two constants are determined by curve-fitting the stress-strain data.

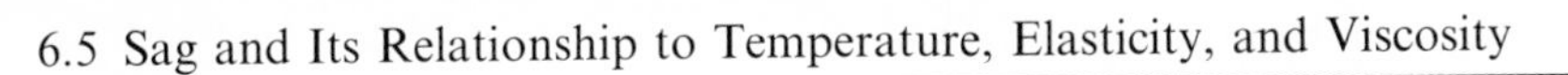

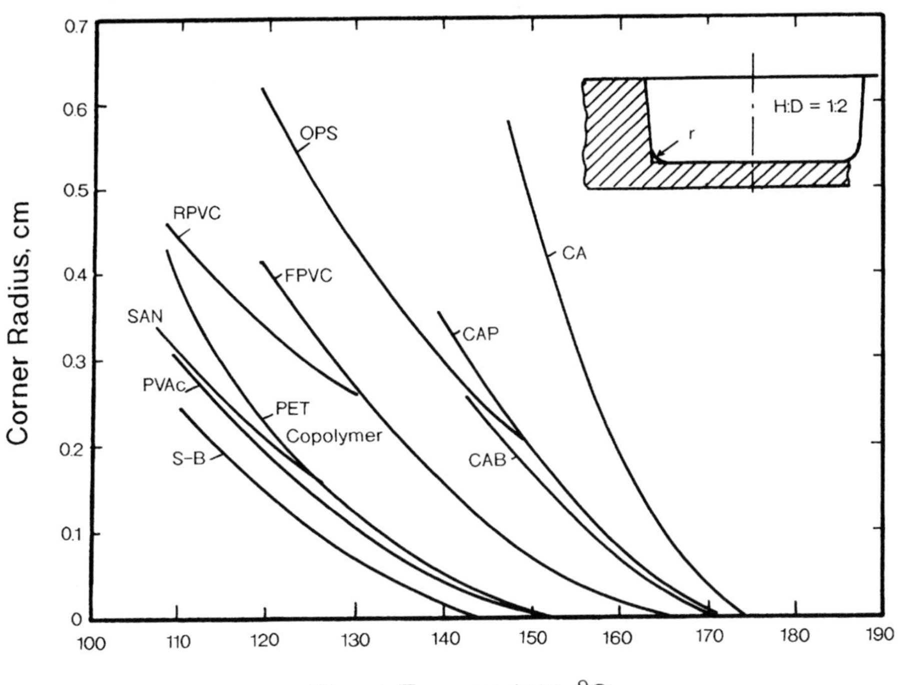

Figure 6.9 Temperature-dependent corner radius for several thermoformable polymers.

Recently, a power-law model, called the Ogden model, has been adopted. Usually either two or four constants are determined by curve-fitting the stress-strain data. Unfortunately, adequate temperature-dependent stress-strain data are lacking for most polymers.

Although elastic stress-strain describes most of the thermoforming process, there is growing evidence that sheet deformation during plug assist pre-stretching is viscoelastic. Design models incorporate the polymer time- and temperature-dependent viscous response through a complex model known as the K-BKZ model. However, again, adequate values for the constants in the K-BKZ model are not available for most polymers.

6.5 Sag and Its Relationship to Temperature, Elasticity, and Viscosity

When a cold polymer sheet is edge-clamped and held horizontally, its weight causes it to sag in the middle. Thicker sheet and longer-dimensioned sheet sag more than thin sheet or shorter-dimensioned sheet. When the sheet is heated, its strength decreases and the amount of sag increases. The rate and extent of sag has been related both to the temperature-dependent elastic modulus of the polymer and its time- and temperature-dependent extensional viscosity.

As seen in Fig. 6.1, the elastic modulus of PC above its glass transition temperature is higher than that for PMMA. PMMA's elastic modulus at its forming temperature is greater than that for ABS. Practically, PC shows very little sag, even at very high forming temperatures, while ABS exhibits substantial sag. Figure 6.10 shows one conceptual correlation between the time-dependent sag rate and time-dependent extensional viscosity. When the extensional viscosity increases, the time-dependent extent of sag decreases, as seen for increasing levels of acrylic modifier in PP in Fig. 6.11.

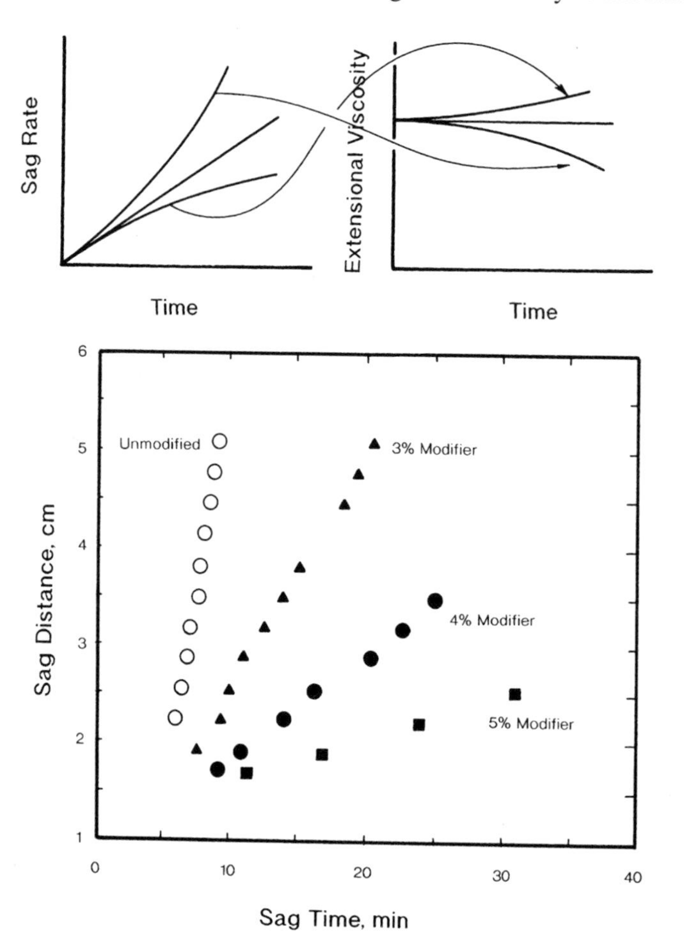

Figure 6.10 Relationship of elongational viscosity and sag rate.

Figure 6.11 Effect of acrylic modifier on sag rate of polypropylene.

6.6 Cooling Against a Mold Surface

Mold material characteristics dictate the way the plastic cools. Mold materials and mold design are discussed in Chapter 8. Metal molds are used for production, with aluminum the predominate material. Cooling channels or cooling plates are used to remove the heat from the formed plastic part. Wood, plaster, syntactic foam, and aluminum are commonly used for prototype tooling.

Usually, non-metallic, prototype tooling does not incorporate cooling channels or plates. For single-surface molds, the majority of the heat in the formed plastic part is removed by the coolant flowing through channels in the mold body. The free surface of the formed part is exposed to ambient air. If fans or blowers are used to circulate the air past the free surface of the formed part, heat removal can be accelerated by 10% or more. When matched tooling is used, both sheet surfaces contact cooled molds. The cooling rate is increased by about a factor of four. For heavy-gauge sheet, the mold surface temperature should be at least about 10 °F (5 °C) below the glass transition temperature, the heat distortion temperature, or the recrystallization temperature of the polymer.

Although low mold surface temperatures decrease cycle times, excessive residual stresses are locked into the formed parts under these conditions. Some of these stresses may be relieved during trimming, when the product is stored at elevated temperatures, during shipment, or in use. These stresses can result in a deformed or distorted part. Mold surface temperatures should be high if replication of the mold surface on the sheet is desired. For example, if a textured sheet is being formed, the sheet's temperature should be low and the mold's surface temperature should be high.

For thin-gauge sheet, the mold surface temperature should be above the temperature at which in-mold condensation can occur. Condensation causes dimples in the walls of formed parts. If the thin-gauge part is to be reheated in use, mold temperatures should be similar to those for the same polymer when it is used for heavy-gauge forming.

Water is the common cooling medium. It is inexpensive, abundant, and is very efficient in removing heat. Water recirculating systems, common in injection molding, are less common in thermoforming, as are elaborate gun-bore drilled water lines and extensive manifolding. In certain instances, where mold temperatures exceed 185 °F or 90 °C, hot oil is used. Electric cartridge heaters are used when very high-performance parts are formed. Mold temperature control is then accomplished indirectly with blowers aimed at the mold surface.

Aluminum has one of the highest thermal conductivities of all common mold materials used in thermoforming. As a result, flooded cooling plates are used extensively in thin-gauge thermoforming and with built-up molds in heavier-gauge thermoforming. Water channels are drilled for some deep thin-gauge multicavity molds. Water lines are cast into or welded out the rear surfaces of heavy-gauge cast aluminum molds. These lines are then connected through manifolds to the external water circulating system through quick disconnects.

To achieve uniform heat transfer across the mold surface, it is imperative that the fluid be turbulent everywhere in the coolant channel, regardless of the fluid used. To achieve turbulence, water should be flowing at least 1 ft/s (0.34 m/s) in 1-in (25 mm) internal diameter lines and 2 ft/s (0.7 m/s) in 1/2-in (13 mm) internal diameter lines. Typically, heat transfer oil has higher viscosity and lower density than water and as a result, must flow faster to remain turbulent.

Temperature across the mold surface should vary by no more than 2 °F (1 °C) and the coolant temperature rise should be no more than 5 °F (3 °C). Part-to-part non-uniformity in thin-gauge multicavity molding and warping and side-wall distortion in heavy-gauge molding are frequently directly attributable to non-uniform mold surface temperature.

6.7 Heat Removal by Mold and Coolant

Typically, there is relatively little increase in mold surface temperature as heat is removed from the sheet during production, even if the coolant lines are some distance from the mold surface. On deep cavities however, the rim regions of the mold surface may increase as much as 10°F or 5°C during production even for relatively thin sheet. Channel cooling is needed if this order of temperature increase is measured. Cooling time frequently controls the overall cycle time in heavy-gauge thermoforming. One reason for this is that both surfaces of the sheet are heated but heat is removed by conduction primarily through only one surface, i.e., the one in contact with the mold surface. Recall that in conduction, the cooling time is proportional to the square of the part thickness.

Three factors mitigate this apparent mismatch in heating and cooling times. First, the part being cooled has been stretched and drawn onto the mold surface. As a result, the part wall thickness is less than the initial sheet thickness. Keep in mind however, for parts with non-uniform wall thickness, the cooling time is dictated by the time needed to cool the thickest part.

Second, the part does not need to be cooled to room temperature, only to the temperature where it is sufficiently rigid to withstand removal from the mold without further distortion. This means that the energy to be removed by the mold may be only 60 to 70% of the energy added during heating.

The third factor is free surface cooling. The free surface is usually cooled by convecting air across it, using fans, blowers, or even water fog or mist. As seen in Table 6.2, forced air is up to ten times more effective in removing heat than is quiescent or still air. A water spray is more than a hundred times more effective.

6.8 Forming Times

Observation of the forming process yields important information about the rate of forming. Initially, it requires a differential pressure of only few lb/in^2 or MPa to deform the rubbery plastic sheet into or onto the mold surface. But, the sheet is cooling rapidly as it is being stretched. As a result, the elongational resistance to the applied load does not follow one of the constant temperature lines of Fig. 6.5. Instead it follows an effective stress-strain curve shown as a dotted line on the figure. Since the sheet resistance to stretching is increasing with time, the observed result is a significant slowdown of the forming process. Typically, most forming takes place early in the process. If the sheet temperature is decreasing rapidly, there may not be enough stress or applied differential pressure to press the sheet into corners with very small radii.

Remedies to this thermoforming version of a "short shot" include forming faster, using higher differential pressure, and heating the sheet to a higher temperature. Thin-gauge thermoforming cooling times increase about 1 second for each 5 mils (0.005 in or 0.13 mm) in sheet thickness. The cooling time for 160 mil (0.160 in or 4 mm) thick ABS is about 60 to 70 seconds. The cooling time for an HDPE part of the same thickness is about 110 to 130 seconds. The difference is the result of the recrystallization heat removal required for HDPE. Forced air on the free surface reduces the cooling time by about 20% in both cases.

7 Trimming

Trimming is usually the process of mechanical breaking one piece comprising the formed part and edge trim into at least two pieces, one, the desired end product and the other, excess material to be reground and reprocessed. Nearly all as-manufactured thermoformed parts must be trimmed in some fashion. The exceptions are products such as inner liners and skylights, where the unformed edges of the part are inserted into assemblies.

There are two types of trimming. The first involves separating the formed part from the web, skeleton, or edge that was used to hold the sheet during heating and forming. The second involves secondary punching or machining of the part itself. The gauge of the sheet dictates the nature of trimming. Thin-gauge trimming and heavy-gauge trimming are considered separately in this chapter.

7.1 General Trimming Concerns

For the most part, trimming is mechanical cutting or breaking of plastic. Two general approaches are used. Compression cutting involves a sharp, toothless blade that is mechanically pressed into the plastic. The second involves a multi-toothed blade that is forced into the plastic at an angle. The individual teeth break small pieces of plastic away, forming a kerf through which the blade follows. Bandsaws and rotary saws are common multi-toothed blades, as are router tips and rotating drill bits. This approach is called chip cutting or fracture cutting.

Surprisingly, all mechanical cutting techniques generate very fine dust or fibers that usually need to be removed from the trim area. Vacuum pick-up at the cutting site aids in dust and fiber control. Certain polymers, such as PS and PMMA, generate trim dust with a high static charge. Antistatic coatings and ionized air directed at the cutting site may mitigate tenacious dust attraction.

7.2 Thin-Gauge Trimming

As noted in Chapter 4, there are three general places for trimming parts in roll-fed thermoforming. In-place trimming combines the forming and trimming functions of the process. These molds are far more complex and more expensive than molds with no trimming function. In-machine trimming incorporates the trimming function downstream of the forming step, but within the thin-gauge thermoforming press frame. Frequently, in-machine trimming includes part-web separating and stacking functions. Stand-alone trimming presses are used for in-line trimming.

7.2.1 In-Place Trimming

In-place trimming offers several advantages (Fig. 7.1). The formed part is locked in place by the forming process. As a result, there is no concern about registering the part prior to trimming. The cutting die easily cuts through the very warm plastic, without substantial generation of trim dust or fibers. The trim die, when advanced only part of the way into the plastic, acts as a grid or cavity isolator. The cutting blade is a steel rule die (Fig. 7.2), usually made of hardened tool steel sharpened to an edge of about 2 to 5 mils (0.002 to 0.005 in or 50 to 250 μm).

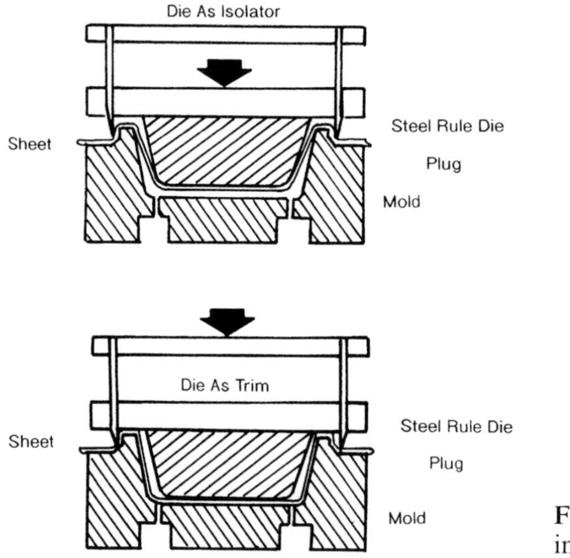

Figure 7.1 Thin-gauge trim-in-place die assembly.

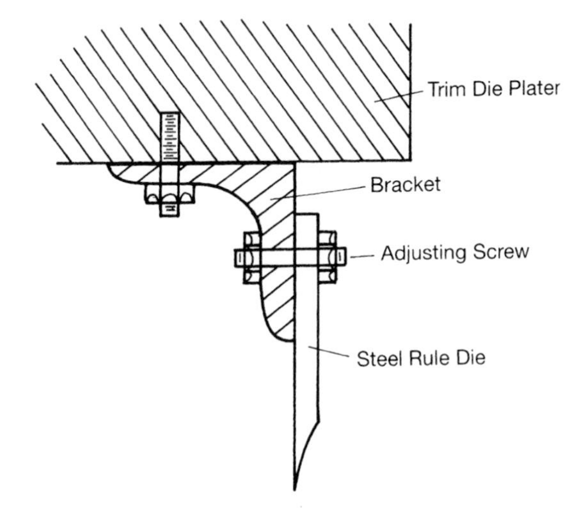

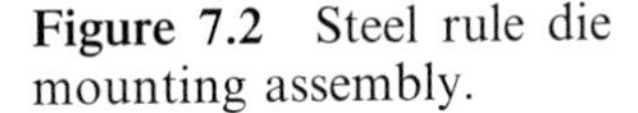
Figure 7.2 Steel rule die mounting assembly.

In some cases, the trim blade passes through the plane of the sheet into a slotted recess (Fig. 7.3). The slot must be wide enough to allow for thermally induced dimensional changes in both the trim die and the mold. If this slot is too wide, warm plastic is sheared between the slot wall and the cutting blade, inducing fiber generation. If machine controls allow very tight trim tolerance, the trim die forces the plastic sheet against a flat metal surface, called an anvil, as shown in Fig. 7.1.

Typically, parts are not completely trimmed from the web, but are instead tabbed. The parts are then punched from the web in a separate stacking operation. Because the trim die and the mold are integral, trim set-up time is very short compared with other trimming techniques. Very high mold cost, the additional on-mold time required for the trim step actuation, difficulties in determining the condition of the steel rule die edge and trim

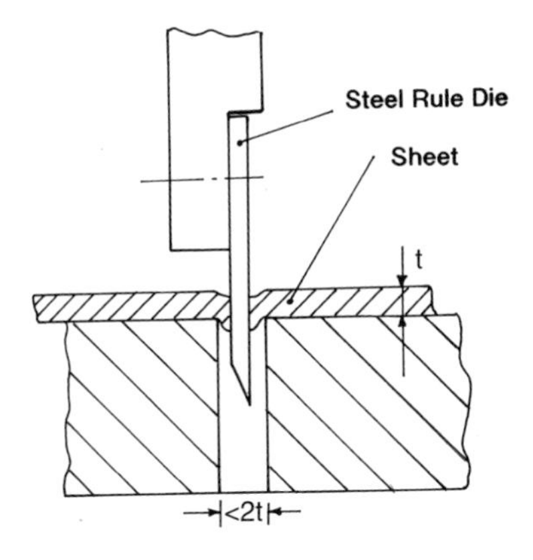

Figure 7.3 Male/Female steel rule die cutting, sometimes called air anvil die cutting.

die alignment, and an occasional part that accidentally separates from the web and remains in the mold are some of the disadvantages to in-place trimming. Part distortion may also be a problem because the plastic is quite warm when it is released from the constraining web. In-place trimming is usually restricted to single-step punching.

7.2.2 In-Machine Trimming

In in-machine trimming, the part is trimmed from the web in a separate station downstream from the mold. Frequently, the trim station also includes part stacking capability (Fig. 4.1). Because the forming and trimming functions are separate, mold costs are lower than those incurred in in-place trimming, and trimming and stacking arrangements are more versatile. Usually, the formed part is fixtured during the trimming step. Trim forces remain relatively low because the part-web is still clamped in the pin-chain and the plastic is still warm from the forming process. The trim and stack sequence may control the process cycle in high-speed forming.

Trim press set-up time may be longer than mold installation and set-up time, particularly if trimming includes stacking. Stacking is usually vertical, either up or down. The punch die can be designed to hole-punch, as well as trim peripherally. But in-line trimming is usually restricted to single-step punching, because the collection and removal of punched-out holes are problematic.

7.2.3 In-Line Trimming

A separate station is usually used for very large volume, dedicated thermoforming operations. In in-line trimming, the canopy, camel-back, or hump-back press (Fig. 7.4) is common. Because the punching action is horizontal, the press is sometimes called a horizontal press. The part-web sheet from the roll-fed thermoformer is directed away from the pin-chain overhead to a sliding arch-type frame that directs the sheet vertically downward into the punch area. The sheet is mechanically registered and indexed so that the formed parts seat in the punch pockets. The trim engine reciprocates the die into the punch, severing the part from the web and pushing the severed part out of the punch to a collection table. Punching rates of 50 cycles are possible. The parts are delivered from the canopy press horizontally, allowing for manual picking and packing.

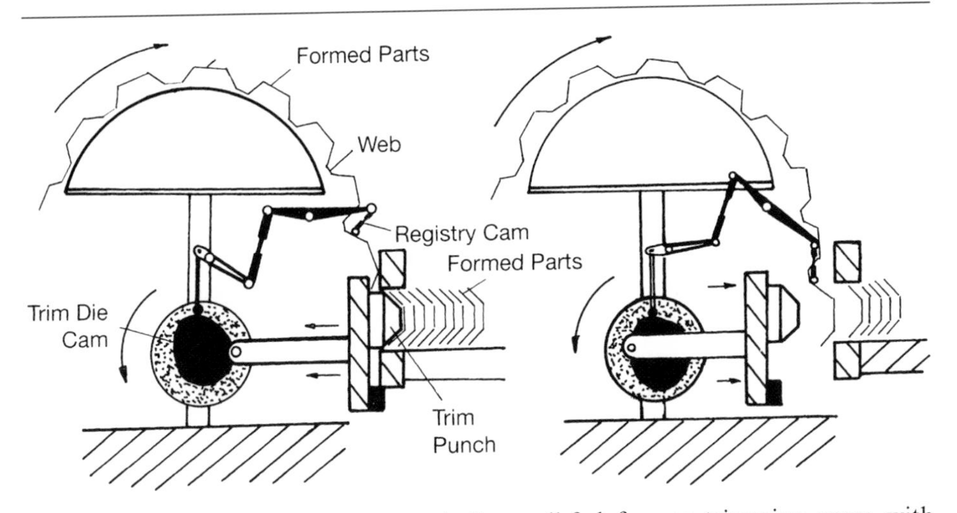

Figure 7.4 Camel-back or canopy in-line roll-fed former trimming press with cam-operated punch-and-die trimmer.

The trim dies are really punches and nesting dies, and are usually forged or machined to tolerance. Canopy machines are notoriously dif-ficult to set up. Deep drawn parts require relatively massive punches and dies. As a result, usually only half a multicavity formed shot is trimmed at one time. Because canopy machines tend to run slowly, punching time may be the limiting part of the process, particularly in high-speed multicavity forming operations.

Tandem canopy presses are used for multiple punching steps for products such as berry boxes and point-of-purchase containers. Although more than one punching station can be built on a single canopy press frame, a single press is not as versatile as two canopy presses in tandem and set-up is about as difficult.

Flat-bed or vertical presses are also popular. The part-web sheet is fed from the thermoforming machine. Registering and indexing pulls the sheet onto the punch and the die activates to punch the part from the web. The parts are either vacuum-picked from the punch face or are stacked by the die pushing the parts through the punch face. The punch-and-die design is quite similar to that used in canopy presses. The collection of punched parts is more difficult than from the canopy presses, but set-up is typically less difficult than with canopy presses, because heavy punches and dies are not subject to gravitational problems. Punch platens can be large enough to trim the entire multicavity formed shot at once. Although multiple punching is feasible on vertical presses, the problems of collecting and removing punch-outs usually dictate tandem presses.

7.2.4 Other Ways of Trimming

There are two common ways of trimming prototype-formed parts. The flat press, shown in Fig. 7.5, sometimes called a dink or clicker press, uses a steel rule die mounted on the upper platen and a tough polymer anvil or pad against the lower platen. Ultra-high molecular weight polyethylene (UHMWPE) is the polymer of choice for the pad. In an even simpler technique, the formed part, containing a trim channel, is manually placed on the lower pad. The steel rule die, attached to a rigid wooden or plywood plate, is then manually placed in the trim channel. The sandwich assembly is then manually run through the nip of the two rollers shown in Fig. 7.6. Of course, hand-held utility knives, linoleum knives, and even scissors are used to trim a few parts.

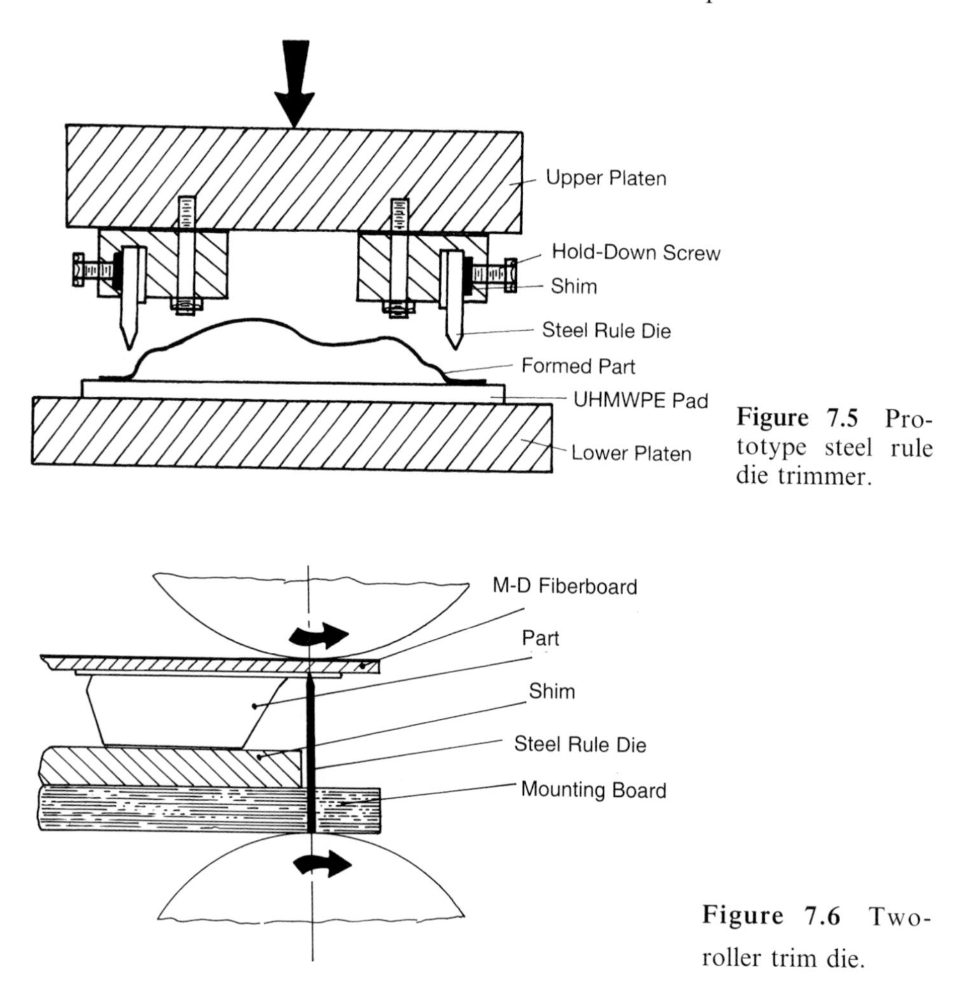

Figure 7.5 Prototype steel rule die trimmer.

Figure 7.6 Two-roller trim die.

7.3 Heavy-Gauge Trimming

Heavy-gauge trimming techniques are divided into in-plane or planar trimmers and shape trimmers. They can also be categorized as manual and automated.

7.3.1 In-Plane Trimmers

In dedicated forming operations, forged or machined dies are built into the forming molds in much the same way as in-place dies are built into thin-gauge molds. These dies are activated while the formed sheet is still in contact with the mold. Guillotines of the type used to cut heavy-gauge sheet are also used in trimming if the part trim lines are planar and linear.

Bandsaws and rotary saws are used if the part trim lines are planar but nonlinear. In many cases, small diameter rotary saws are mounted in custom-designed fixtures and the formed sheet to be trimmed is hand-fed against stops on the fixtures. The typical cutting speed is 1000 to 4000 rpm. The fracture nature of the plastic plays a great role in selecting the right type of cutting tooth, the tooth spacing, rake and kerf, and the vertical speed of the teeth through the plastic. Care must be taken to avoid excessively chipping the edge of the formed part or thermal damage to the plastic. Thermal damage is minimized by using coarse-toothed blades, small blades, low linear velocity, and high feed rates. Wide-tooth spacing is desirable with easily softened plastics such as HDPE and PP. Narrow-tooth spacing is desirable with brittle polymers such as PS, ABS, PMMA, and rigid PVC. Smooth cuts are usually obtained with hollow ground blades with no tooth set or with carbide blades.

7.3.2 Non-Planar Trimmers

Hand-held routers, such as the industrial Dremel router, have been used for decades for trimming. Usually, the part is placed on a fixture that contains grooves to accommodate the router tip. A router path should be formed into the part to facilitate tracking. Because electric and air-driven hand-held routers operate at 12,000 to 25,000 rpm, the router tip must be specially designed for trimming plastics.

Conventional wood rasp ball or cone ends quickly overheat and fuse plastic into the cutter grooves. As with toothed saws, the router must move at a controlled speed to minimize overheating and excessive microcracking. Vacuum fixtures are used with hand trimming as well as with the computer-driven multi-axis machines described below.

7.3.3 Milling Machines

Thermoforming has replaced injection molding as the preferred means of manufacturing quality components for relatively small production runs of high-performance devices. Many of these sorts of parts require multiple holes, slots, vents, cut-outs, and locally accurate wall thicknesses, as well as very accurate trim line locations.

It is well known that a person operating a hand-held router can trim a plastic part more rapidly than current automatic multi-axis routers. It is also well known that an automatic router trims each part to the exact dimensions as the previous one and never exhibits arm weariness, inattention, or tiredness. Common metal milling machines are in fact, three-axis or X-Y-Z machines. The part is held in a fixture that is moved in a horizontal or X-Y plane, and the milling head is moved vertically in the Z plane. The positional accuracy on these machines is typically 1 mil (0.001-in or 25 μm) or better. Most commercial milling machines are numerically or NC-controlled. Drilling and milling sequences and tool paths are programmed into the machine database for each part design. Small plastic parts are frequently trimmed and holes and slots are machined-in on these three-axis machines. Because the movement of the milling head is restricted to a single direction, complex trim lines are difficult. Complex to-be-trimmed parts need to be frequently repositioned on the X-Y platforms. The linear milling speed of typical milling machines is about 100 in/min (2.5 m/min), and cutting head speeds are on the order of 3000 rpm.

7.3.4 Multi-Axis Routers

Five- and six-axis routers have been designed specifically for thermoforming. As with the three-axis machine, the part is moved horizontally in the X-Y plane. The engine for the router head is mounted on an overhead gantry, allowing it to move vertically in the Z direction. In addition, the engine is gimbaled to allow the cutter to move in either two (U and V) or three (U, V, and W) directions. This operation is best envisioned by rotating one's wrist.

The two or three additional degrees of freedom allow the cutter tip to move diagonally or horizontally into the part, and even cut underneath an overhanging portion of the part, while the part is affixed to the X-Y platform. Initially, these special purpose robots were very expensive, relatively slow, difficult to program, and prone to gantry ringing or vibration. Typical accuracies of these earlier machines were about 0.005-in (5 mils or 250 μm). Competition has now brought down the price of these machines. In addition, engine movement is now driven by linear motors rather than spiral screw-driven motors. Motor weights are dramatically lower, thus minimizing inertial effects when the motor position is stopped and started. Computers now control acceleration and deceleration, again minimizing vibration. And, most importantly, linear travel speeds have increased to 1000 in/min (25 m/min) or more and router tip speeds are 40,000 rpm. Newer machines claim accuracies of about 0.002 in (2 mils or 50 μm).

Even with these advances, computer-aided trimming machines have much longer cycle times than the thermoforming machines molding the parts to be trimmed. It is believed that to achieve cycle time parity between the forming step and the trimming step, linear trimming speeds need to be 5000 in/min (130 m/min), router tip speeds need to reach 70,000 to 100,000 rpm, and tool-path programming needs to be far faster and more user-friendly.

7.3.5 Cutter Design

Probably the greatest advancement in automatic trimming is in cutter design. As linear and rotary trimming speeds increase, old cutter designs have become inadequate. The cutting surface must be able to separate a small portion of plastic from another portion without melting the two, generating fine dust or fibers, or causing microcracks. It must also be able to toss that small piece of plastic from the cutting arena before separating a second piece. Even now, cutters are being designed for very specific types of polymers. For example, single-edged and double-edged spiral-out flute cutters are recommended for harder plastics such as ABS.

8 Molds and Mold Design

The thermoforming mold serves several purposes. First and foremost, it must accurately allow production of a part that meets all the customer's specifications. Then, it must provide a dimensionally stable surface against which the formable plastic sheet is pressed. It must be a heat exchanger, removing heat from the sheet in a repeatable and efficient way. It frequently is a pressure vessel. It must allow residual cavity air to be removed through the mold in a controllable fashion. It must be robust enough to withstand repeated formings at elevated pressures and polymer sheet temperatures, in the presence of possibly corrosive gases from the plastic, despite erosion and wear from filled or reinforced plastic, and despite various environmental conditions during long-term storage. It frequently is a machine with manually or automatically actuated cams and slides. It may include plug platens, isolation grid platens, ejection platens, and in-place trim dies and anvils. There are two general categories of thermoforming molds. Production molds are typically aluminum. Prototype molds can also be made from metal, but are usually fabricated of more easily worked materials.

8.1 Production Mold Materials

Aluminum is the standard metal used for thermoforming molds. Atmospheric or foundry cast aluminum is usually selected for large surface molds. Aluminum alloy A-356.2 has excellent castability, machinability, and weldability. Sand is usually used for the pattern. Care must be taken when casting aluminum to minimize porosity and locally soft areas near the surface of the casting. Ceramic patterns and pressure casting are used for pressure forming molds, detailed molds, and molds requiring extensive post-casting machining or finishing. Typically, the casting is nearly uniform in thickness at about 1 in (25 mm) or so. The completed casting is deflashed and risers are machined away. Coolant lines are then welded or soldered against the back of the casting, vent holes are drilled in, and positive surfaces are finished.

With the development of computer-aided milling machines, machined aluminum plate is a standard way of manufacturing multiple cavity molds. Al 6061-T or aircraft aluminum is commonly used. Al 7075-T, a tougher, higher temperature aluminum, is sometimes used for polycarbonate and certain filled and reinforced polymers. Cooling channels are gun-bore drilled for deep-draw molds. For shallow molds, a cooling plate is mounted to the mold base.

Steel is sometimes used for high-temperature polymers. Prehardened P20 steel is recommended for molds that require high pressure, temperature, and wear resistance. Electroformed nickel is used for very large, highly detailed parts. Cold electroforming against a conductive pattern yields the best surface quality. The desired nickel thickness is 60 mils (0.060 in or 1.5 mm). After the coolant lines are in place, the nickel is then backed with hot electroformed metal, cast aluminum, sprayed white metal, or aluminum-filled epoxy.

8.2 Prototype Mold Materials

Because the pressure in traditional vacuum forming is low, many common materials are used as mold materials. Two general techniques are used to construct prototype molds. Deductive manufacture is the removal of material to create the desired surface. Inductive manufacture is the build-up of material to create the desired surface. Hardwoods, industrial plaster, fiberboard, syntactic foam, thermoset plastic, and sprayed white metals backed with epoxy are materials of choice. Hardwoods such as hard maple, hickory, and ash are easily shaped, drilled, and sanded and so are used for very short runs of a few parts. The common problem with wood is excessive drying that leads to splitting, checking, and warping. For this reason, wood should be used without post-applied coatings or finishes such as epoxy or polyurethane. The compressive strengths of most woods parallel to the grain are on the order of 4000 lb/in^2 (27 MPa). This value restricts wood's use to vacuum or very low pressure forming. Wood working is deductive.

There are many grades of industrial plaster. Plasters with compressive strengths of at least 5000 lb/in^2 (34 MPa) are used. Plasters with very high compressive strengths can be used in moderate pressure forming applications. Thin sections are subject to brittle fracture. Sometimes sisal, glass fibers, or hemp mats are used to improve the bending strengths of the mold. Plaster mold manufacture is inductive until the mold is nearly finished. Plaster is then removed to finish surfaces to tolerance and to add necessary vent holes.

Medium-density fiberboard (MDF) is used to make shallow-draw male molds. MDF is a pressed wood fiber product that can be sawed, shaped, and drilled with common wood-working tools. Because MDF is semi-porous, vacuum can be drawn through the body of the mold without drilling vacuum holes (Fig. 8.1). The compressive strength of MDF is less than 5000 lb/in^2 (34 MPa). As a result, it is not recommended for pressure forming. The surface of an MDF mold is matte.

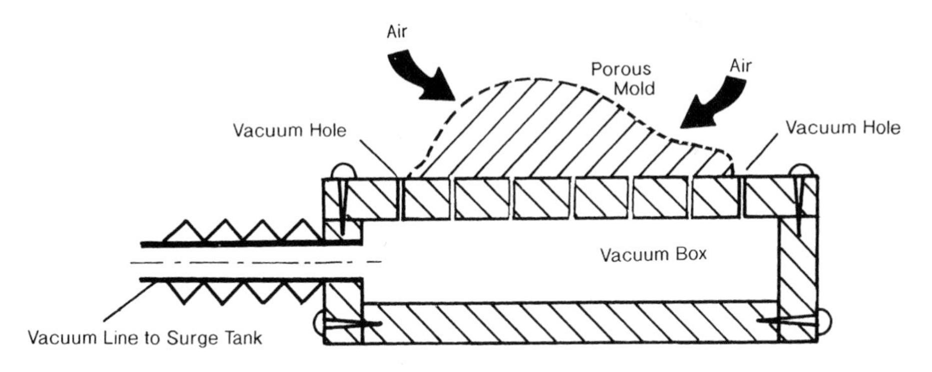

Figure 8.1 Prototype mold assembly for porous, medium-density fiberboard or porous aluminium.

Syntactic foam is made by incorporating sintered or foamed micro-spheres of fly-ash; phenolic; or hollow glass in polyurethane, phenolic, or epoxy resin. The polymer may or may not be foamed. Syntactic foams are usually available as plank or rod stock in densities of 12 to 50 lb/ft^3 or 200 to 800 kg/m^3. Castable versions are also available. They are easily machined and drilled, using carbide-tipped machining surfaces. The compressive strength of commercial syntactic foams is about 6500 lb/in^2 (44 MPa). Syntactic foams are commonly used as plugs. They are not usually used in pressure forming applications.

Thermoset plastics are used in prototype forming when a few very large area parts are needed. Glass-reinforced epoxy and high-temperature ther-mosetting polyesters are materials of choice. Compressive strengths in excess of 10,000 lb/in^2 (68 MPa) are achievable. Continuous use temperatures are restricted to 260 °F (125 °C) to minimize thermal degradation of the plastic. Prototype plastic molds are used in multilayer and composite forming.

White metals such as zinc and zinc alloys can be sprayed as molten drops. Spray metal molds can be made in a few hours, include coolant lines (Fig. 8.2), use the actual part as a pattern, and yield surfaces with extreme detail. Despite these advantages, sprayed metal technology has not

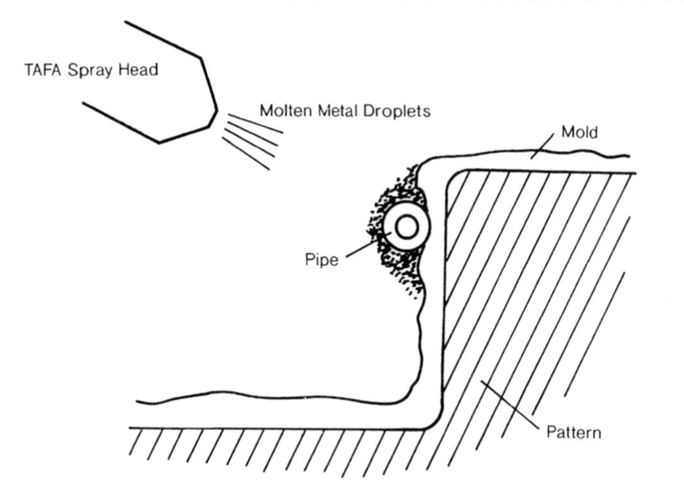

Figure 8.2 Method of fastening reinforcing pipe or coolant channel to back of sprayed zinc metal prototype mold.

attracted prototype thermoform mold makers. Sprayed metal mold shells are typically 0.25 in (6 mm) thick. Because zinc and its alloys are relatively soft, the mold shells are typically backed with aluminum-filled epoxy. As with all polymer-backed mold surfaces, continuous operating temperatures are restricted to 260 °F (125 °C).

8.3 Mold Design

Cooling, venting, undercuts, and surface texture are important elements of mold design. Details about these and other aspects of mold design, such as coining, rim rolling, and plug design are discussed here.

8.3.1 Cooling

The importance of cooling channels is discussed in Chapter 6. Technically, the best coolant pattern employs coolant channels everywhere just beneath the mold surface. Practically, this is not common practice. The high thermal conductivity of aluminum allows coolant lines to be some distance from the mold surface. In certain cases, molds may contain no coolant lines. Instead, the molds are mounted directly against flooded cooling plates. Regardless of the design of the cooling channel, the key is uniform mold surface temperature during commercial forming. Manifolding is always recommended over serpentining. Sufficient cooling paths are needed to maintain a coolant temperature rise of no more than 5 °F (3 °C).

8.3.2 Venting

Quality parts are made on molds that have controllable air evacuation systems. The evacuation system begins with vacuum or vent holes drilled through the primary mold surface. These vacuum holes are connected to machined vacuum channels. These in turn are connected to a plenum or vacuum box (Fig. 4.7). The vacuum box is connected to the vacuum pump through a vacuum line. A solenoid-activated rotary valve and a surge tank are also in the line.

All molds need adequate venting to quickly remove the mold cavity air trapped between the mold surface and the deforming sheet. The number of vacuum or vent holes is dictated by the rate of air evacuation from the mold cavity. It is imperative that the stretching of the sheet not be restricted by a cushion of air in the mold.

Plastic is stretched last into the three-dimensional and horizontal two-dimensional corners of a mold. As a result, vacuum holes are always put there. Vacuum holes are also put along vertical two-dimensional corners, vertical surfaces, and the lip and rim areas of the mold. Vacuum holes are usually incorporated in raised regions such as logos, denesting lugs, and partitions. Vacuum holes are usually spaced very regularly along corners primarily for esthetic reasons. Spring-loaded valves connected to auxiliary evacuation devices are used to evacuate very deep molds.

The diameter of any vacuum hole should not exceed the thickness of the sheet that covers it. If the vacuum hole is too large, the hot plastic sheet is thermoformed into it. The resulting nibs or nipples on the formed part may be undesirable. The smallest vacuum hole commonly drilled is #80, or 13.5 mils (0.0135 in or 340 μm) in diameter. Porous sintered metals with micron-sized pores are used when smaller vacuum holes are needed.

8.3.3 Undercuts

Because thermoforming is a low pressure, low temperature, single-surfaced process, it is ideal for the production of parts with severe undercuts. Undercuts are common features on both thin-gauge and heavy-gauge parts. Many thin-gauge parts with undercuts are just pulled from the mold surface, the plastic simply bending to slide free. When necessary, ejection or stripper elements are used. When the undercut is severe, the polymer is stiff or brittle, or the formed part is very thick, the mold must move to allow the release of the part. The movable section can be simply hinged to the rest of the mold (Fig. 8.3), or it can be pneumatically moved (Fig.8.4). In extreme cases, the part can be molded around an element that is removed with the part, then reassembled in the mold for the next cycle.

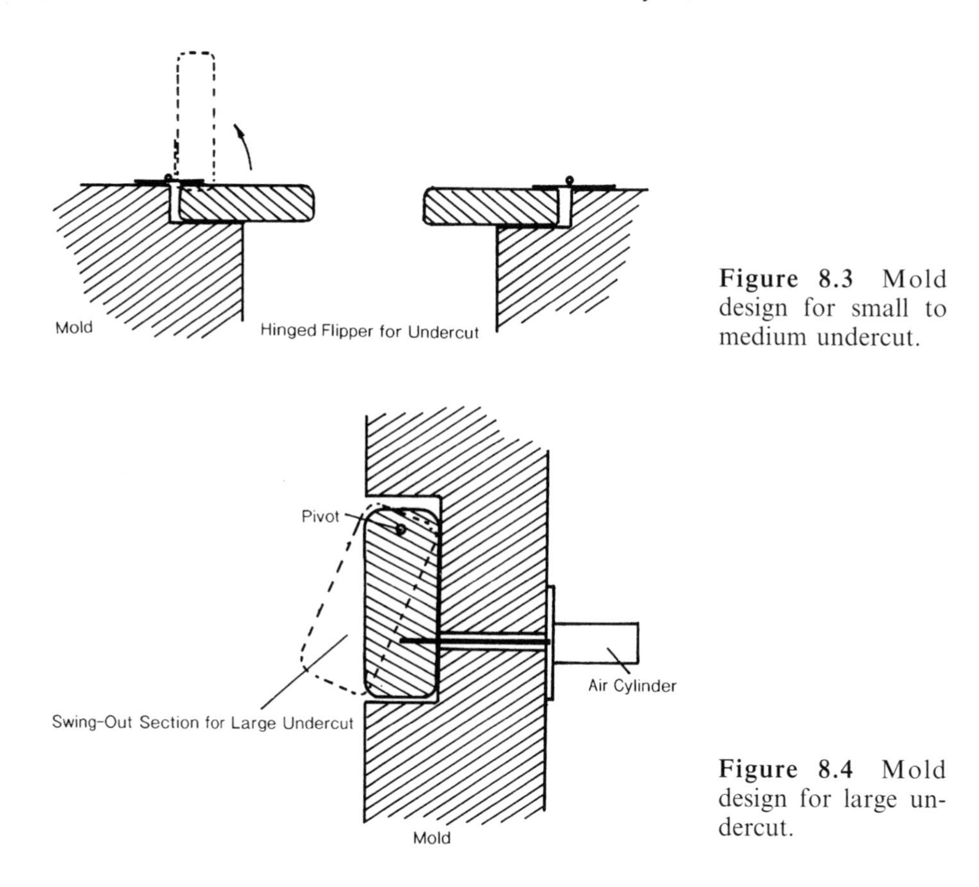

Figure 8.3 Mold design for small to medium undercut.

Figure 8.4 Mold design for large undercut.

8.3.4 Mold Surface Texture

The hot plastic sheet does not replicate fine details on the mold surface in traditional vacuum forming. The low pressure and the rubbery, solid nature of the polymer prevent the sheet from picking up mold surface details smaller than about 2 mil (0.002 in or 50 μm) in height. Nevertheless, whenever a hot plastic sheet touches the mold surface, it deglosses to a matte finish. If a part must have a glossy surface, it can be formed free of a mold, as with freely blown skylights. Or, the surface can be flame treated or polished with 2400 grit Crocus cloth.

For very hot, rubbery solid, or elastic liquid polymers, and certainly for pressure forming, mold surface textures as fine as 0.4 mils (0.0004 in or 10 μm) can be replicated. Sixty to 100 mesh grit or sand blasted mold surfaces typically have roughness dimensions of 8 to 10 mils (0.008 to 0.010 in or 200 to 300 μm). Chemically etched mold surfaces may have roughness dimensions down to 0.5 to 5 mils (0.0005 to 0.005 in or 13 to 130 μm). The very fine texture of these mold surfaces can be accurately and reliably replicated only by pressure-formed hot sheet surfaces.

8.3.5 Textured Mold or Textured Sheet?

Mold surface texture and smooth sheet surface is the common combination for forming a part with a textured surface. However, there are occasions when a smooth mold surface and a textured sheet surface is desired. As one example, the appearance surface must be the free surface. In another, the mold cannot be textured, as is frequently the case for a prototype mold. In a third, the mold design is such that the desired texture cannot be applied uniformly over the entire mold surface.

8.3.6 Plug Assist Materials and Designs

Syntactic foams, solid plastics such as nylon and UHMWPE, heated aluminum, and felt-covered wood are used as prestretching plugs. The material choice depends on whether the operation is production or prototype; the plug design is still undergoing changes; the polymer sheet is marked by certain plug materials; the plug temperature needs to be carefully regulated; the sheet is thick; or the sheet is hot.

Wooden plugs are used in most prototyping operations and in many heavy-gauge productions. Wood is light in weight, easily manufactured and reworked, and has low thermal conductivity. The wood surface is usually covered with felt. Syntactic foams are developed specifically for plugs for high-performance, thin-gauge thermoforming. In certain instances, the foam surface is coated with polyurethane, epoxy, or Teflon* (PTFE or FEP). Temperature-controlled machined aluminum plugs are used for polymers such as crystallizing PET (CPET) and oriented polystyrene (OPS).

Rules for plug design are not well documented, but certain guidelines are known. The plug surface should not mark off the part, either by chilling the sheet too quickly or transferring its texture to the sheet. Several plug shapes

*Teflon is a registered trademark of E.I. duPont de Nemours and Co., Inc.

are given in Fig. 8.5. The plug shape is dictated by the amount and location of material redistribution. Flat-bottomed plugs are used if substantial side-wall stretching is needed. Bull-nosed plugs are used to redistribute polymer from both the bottom and side-walls of the part. Ring plugs are used if the bottom of the sheet must stay hot or if the sheet must stretch over a center column. Articulated plugs are used if the sheet must be tucked under an overhanging portion of the mold. Tapered plugs are used when the mold is tapered. Typically, the plug should be about 80% of the mold dimension and should penetrate to about 80% of the depth of the mold cavity.

8.3.7 Other Mold Features

In certain instances, the polymer must be locked against the rim prior to plug-assist forming. Ridges, called dams, or grooves, called moats, are used

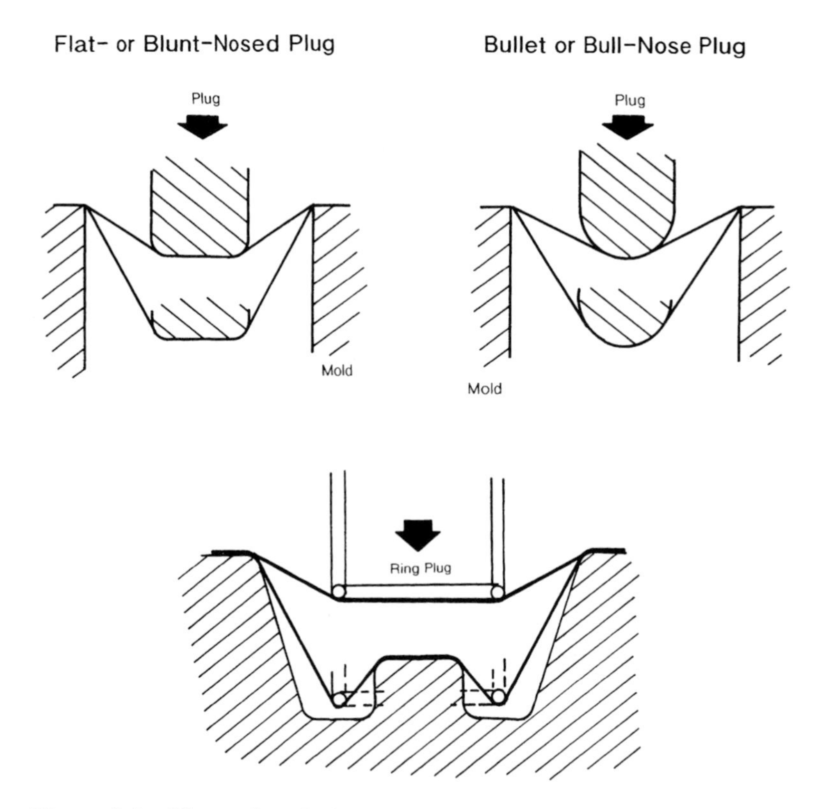

Figure 8.5 Three plug designs.

for this purpose. Chamfers are recommended instead of radiuses when three-dimensional corners are too thin, even with zonal heating and prestretching. Coining is the localized pressing of the sheet while in the mold. Figure 8.6 illustrates coining for either a logo or other identification. Coining is also used to produce a specific thickness in a given area of the part, possibly for assembly purposes, and to produce a dimensionally accurate flange in thin-gauge forming for hot sealing purposes.

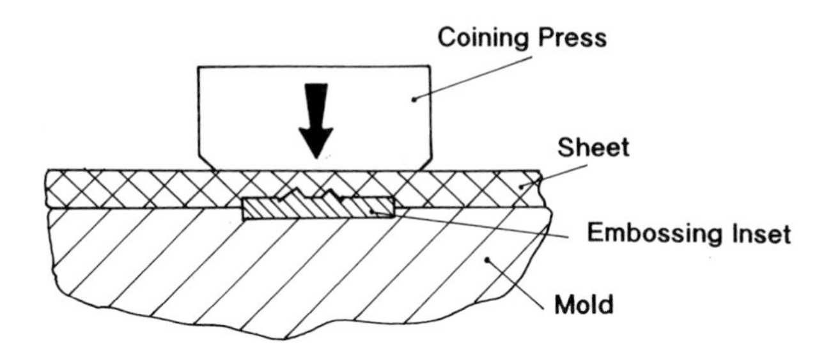

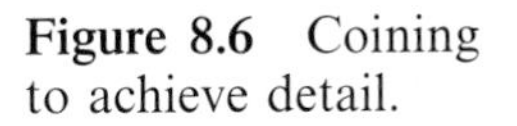

Figure 8.6 Coining to achieve detail.

8.3.8 Molds for Twin-Sheet Forming

Twin-sheet forming requires additional mold features. For simultaneous twin-sheet forming, both sheets are presented to the open mold halves at the same time. Blow pins are placed between the sheets during the loading step. Air is introduced between the sheets during the heating step to keep the sheets separate. During the forming step additional blow pins or needles may be mechanically or pneumatically driven through one of the sheet surfaces to aid in pressing the sheets against their respective mold surfaces. The mold halves close onto the sheets and are clamped to ensure accurate mating of the peripheral seal area. The molds must close with enough force to squeeze the plastic enough to result in fusion. To achieve a strong seal, some of the plastic should be squeezed from the compression area. Seal area designs depend on available clamp force as well as the required appearance of that region on the final part (Fig. 8.7).

In sequential twin-sheet forming, the first sheet is formed against the lower mold, the second sheet is then formed against the upper mold, and the two mold halves are then clamped around requisite blow pins. The first sheet must be dropped from the clamp frame after forming, so that the second sheet can be indexed. Combination moats and dams and even auxiliary clamps are built into the lower mold to keep the first sheet from moving, distorting, or shrinking while the second sheet is being formed.

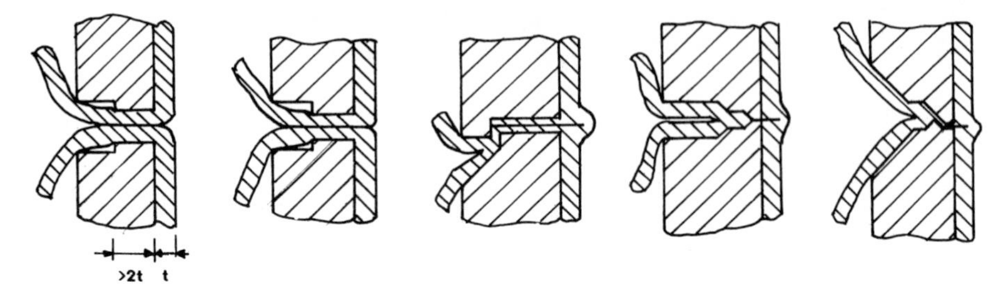

Figure 8.7 Cross-sections of twin-sheet sealing area. First two show effect of increasing pressure.

Auxiliary rod heaters may be built into the lower mold peripheral seal area to keep the sheet sufficiently hot to result in adequate sealing. Sequential twin-sheet forming is used if a stiffening insert is to be encapsulated. Although the insert may be manually added, robotic placement is used if very accurate placement is needed.

9 Part Design

The objective of any thermoforming operation is to produce saleable parts. A wide selection of polymers can be thermoformed and sheet can be formed from the selected polymer by a large number of means. As a result, only general guidelines on part design can be outlined here. First, the general elements of part design are stated. It must be determined at this point whether thermoforming is the best way to make the part. Then, certain limitations of the process must be factored into the decision to help refine the design. As part dimensions become more critical, polymer shrinkage and general dimensional tolerance become more important. Toward the end of this chapter, recent developments in computer-aided *a priori* determination of wall thickness are discussed.

9.1 Part Design Philosophy

At least three concerns must be addressed in determining how to manufacture any product:

- Will the finished part meet all required and specified design criteria?
- Can the part be produced at the minimum cost for the projected market size?
- What are the consequences if the part fails to meet minimum requirements?

9.1.1 Project Protocol

No part design and fabrication program should begin without a strict, formal, written protocol. The protocol begins with a clearly stated, clearly understood project objective and an ancillary list of requirements for product performance. Initial meetings should be face-to-face among all key principals, including representatives from at least one resin supplier, an extrusion house, a mold maker, a machinery builder if a new machine is contemplated, a trimming device supplier if new trimming technologies are needed, a thermoformer judged capable of manufacturing the proposed part, the customer's technical and marketing departments and, in some cases, even representatives from the customer's customer. Additional communication should be by teleconferencing and video-teleconferencing. All principals should be informed of all major developments, including updated market information, important cost revisions, redesigns around molding problems, and the ubiquitous "improvement" modifications.

Material selection should begin by listing those polymer properties needed to meet the performance criteria of the part. A checklist should include topics such as environmental conditions, including extreme conditions; mechanical behavior and tolerances under environmental conditions; dimensional tolerances; and other material specifications such as color fastness and cost.

9.1.2 Should This Part Be Thermoformed?

There are many ways to manufacture any plastic part. The nature of the forming process should be researched early in the evolving protocol. Some reasons for not selecting thermoforming include:

- The inability to satisfactorily extrude the candidate polymer into sheet
- The inability to heat the sheet to a forming temperature without excessive sag
- The inability to stretch the hot sheet into the desired shape
- The unavailability of suitably sized forming machines
- A market that is too small or too large for thermoforming
- The inability to regrind, recycle, or reuse the trim or web
- Part performance that requires highly reinforced polymers
- Part tolerances and draft angles that are unacceptably tight for thermoforming
- The need for parts with highly uniform wall thickness
- The fact that other processes may be more economically competitive

Conservative product design focuses on minimizing the amount of trim and out-of-spec parts that must be recycled. The amount of recycled materials is minimized by selecting:

- Conservative designs rather than exotic ones
- Simple processes, such as vacuum or drape forming, rather than multi-step or unproven forming processes
- Polymers that draw well and are not near their elongational limits
- Sheet that is not too large or too small for the job
- Machinery that is rugged, well-maintained, and adequately controlled
- Workers who are fully trained on all aspects of the forming process
- Incoming materials quality control guidelines that function
- Manufacturing deadlines that are reasonable

9.2 Shrinkage, Draft, Thermal Expansion, and Dimensional Tolerance

Shrinkage, draft, thermal expansion, and dimensional tolerance are probably the most difficult concepts to manage in product design. Shrinkage is a polymer material property. When any plastic is heated, the spacing between its molecules increases. The result is an increase in volume and decrease in density. A crystalline polymer exhibits a substantial increase in volume and decrease in density as it is heated through its melting point. There are no similar jumps in these values when a plastic is heated through its glass transition temperature. When the plastic temperature is lowered, its density increases and its volume decreases.

Shrinkage is always associated with cooling. There are two general types of shrinkage. When any plastic is allowed to slowly cool without being constrained against a mold surface, for example, it shrinks uniformly. The shrinkage is the same in any dimension and is approximately one-third that of the polymer's change in density or volume. The final plastic volume is essentially the same as it would be if it had not been heated and cooled. However, if the plastic is held rigidly against a mold surface, it does not necessarily shrink uniformly in all directions. The more quickly the plastic is cooled, the less likely the polymer molecules will return completely to their initial state.

Constrained shrinkage is an important factor when predicting or determining the causes of gross product problems, such as warp and distortion, and more subtle problems, such as part-to-part variations in dimensions. Non-uniform orientation in the incoming sheet also causes non-uniform shrinkage. Typically, machine direction orientation is greater than cross-machine or transverse direction orientation. For heavy-gauge sheet, the difference in shrinkage values can be as much as 0.2%. Table 9.1

Table 9.1 Shrinkage Values for Thermoformable Polymers

Polymer	Shrinkage Range (%)	Recommended Value (%)
ABS	0.5–0.9	0.7
EVA	0.3–0.8	0.6
FEP fluoropolymer	1.5–4.5	3.0
Polycarbonate	0.5–0.7	0.6
LDPE	1.5–4.5	3.0
HDPE	2.0–4.5	2.5
PMMA	0.2–0.8	0.6
PP	1.0–2.5	2.0
PS	0.5–0.8	0.6
Rigid PVC	0.1–0.5	0.3
K-Resin	0.4–0.8	0.6
APET	0.3–0.6	0.5
CPET	10–18	12

gives representative shrinkage ranges and recommended constrained values for several thermoformable polymers.

Polymers shrink onto male portions of molds and away from female mold surfaces. Draft is required to free cool parts from molds. Typical female mold draft angles are 0 to 2°, with an average of about 1°. Female mold draft angles for crystalline polymers are less than those for amorphous polymers.

Typical male mold draft angles are 1 to 5°, with an average of about 4°. Male mold draft angles for crystalline polymers should be greater than those for amorphous polymers. Typically, draft angles are increased 1° for every 0.2 mils (0.0002 in or 5 μm) of texture depth.

Characteristically, plastics expand and contract at rates five to ten times that of metals. Linear expansion is important when the plastic part is an element of an assembly of other, non-plastic, parts. The linear expansion values in Table 9.2 are used when the polymer is not undergoing melting or glass transition. Filled and reinforced polymers do not expand or contract as much as unfilled and unreinforced polymers.

Part-to-part dimensional consistency is always more difficult to achieve with single-surface molding processes such as thermoforming, blow molding, and rotational molding. In addition to polymer-specific shrinkage and thermal contraction, thermoformed parts are fabricated at relatively low temperatures and very low pressures from rubbery, elastic sheet. As a result, non-uniformity in sheet orientation and variations in the heating,

Table 9.2 Coeffiicents of Thermal Expansion for Thermoformable Polymers

Polymer	Range (10^{-6}/°F)	Range (10^{-6}/°C)
ABS	60–130	35–70
EVA	80–200	45–110
FEP fluoropolymer	35–70	20–40
Polycarbonate	70	40
LDPE	100–220	55–120
HDPE	60–110	35–60
PMMA	50–90	30–50
PP	80–100	45–55
PS	50–80	30–45
Rigid PVC	70	40
K-Resin	65–70	35–40
APET	65	35

stretching, and cooling steps can dramatically influence the final part dimensions.

There are at least two aspects to dimensional tolerance. Local wall thickness variation is strongly related to the geometry of the part and to methods, such as zonal or pattern heating and prestretching, used to redistribute materials across the part surface. There are many processing elements that affect variation in local wall thickness, including incoming sheet orientation and thickness variation, heater cycling, variation in overall sheet temperature at forming time, and mold temperature cycling. Normal wall thickness variation for most traditional thermoforming processes is about 20%. Wall thickness variation for close tolerance parts should be no more than 10%. Figure 9.1 shows wall thickness variations on two consecutive days on a production run of heavy-gauge cabinets.

Because of non-uniform wall thickness, every formed part has a non-uniform stress field. If the plastic is cooled very quickly, these stresses are frozen in place. Non-uniform mold temperature and non-uniform sheet temperature amplify the non-uniform stress field. The result is long-term, unconstrained, non-uniform shrinkage that may appear as distortion, warpage, or wide dimensional tolerance in the final part. If the local wall thicknesses vary from part to part, the local stress fields vary as well. This further acts to distort, warp, and cause part-to-part dimensional variations.

Sheet thickness tolerance should be about 1 to 2 mils (0.001 to 0.002 in or 25 to 50 μm) for thin-gauge sheet, and about 5% for medium- and heavy-gauge sheet. This tolerance must be doubled and added to the dimensional tolerance for inside dimensions on parts formed in female molds.

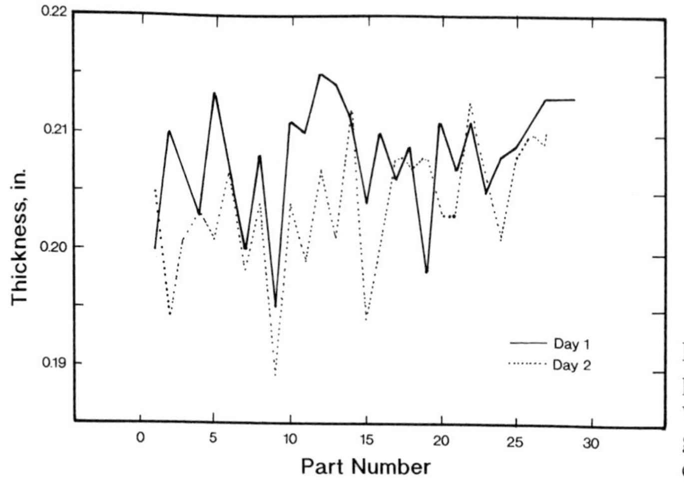

Figure 9.1 Part-to-part wall thickness variation on heavy-gauge thermoformed part.

9.2.1 Improving Dimensional Tolerance

What can be done to improve dimensional tolerance?

1. Incoming sheet characteristics, including thickness and orientation, must be measured and tightly controlled.
2. The forming process, particularly the heating and preforming steps and the mold temperature must also be tightly controlled.
3. Care must be taken to ensure that differential stretching forces, whether simply vacuum or pressure and vacuum, are the same time after time.
4. Simple steps, such as including clamping grids or cavity isolators for thin-gauge multicavity molds, can dramatically reduce part-to-part dimensional variation.
5. A high mold temperature allows stress relaxation and shrinkage to occur while the part is still fixtured, minimizing distortion and warping.
6. Post-mold fixturing aids dimensional stability in parts made of polymers that are slow to shrink, such as HDPE and PP.
7. If long-term distortion is a problem, the parts should be trimmed either earlier or later, depending on the origin of the distortion.
8. Annealing, long used in injection molds, is useful for heavy-gauge fixtured parts.
9. In pressure forming, the sheet is pressed tightly against the mold surface, aiding dimensional control, particularly in female molds.

9.3 Part Wall Thickness Prediction

It is now possible to predict local wall thickness for complex parts from non-uniformly heated sheet that has been plug-assisted through computer software that uses finite-element analysis (FEA). The sheet is electronically replaced with a two-dimensional mesh of triangularly connected nodes. A small force is electronically applied to the mesh, and force balances are made at each node to determine the distortion of the mesh. The force is then increased and calculations are repeated. When nodes touch the electronic boundaries of the mold, they are fixed there. Force is increased until all the nodes touch the electronic mold surface or until the force reaches a preset value.

Intermediate steps illustrate how the sheet deforms against the mold surface (Fig. 9.2). Figure 9.3 compares measured wall thicknesses for several polymers with the computer-generated wall thickness for a thermoformed five-sided box. FEA has replaced geometric analysis, where the sheet thickness is determined everywhere by manually applying equations to increases in surface area as the sheet is stretched into simple geometries such as two-dimensional corners and wedges and three-dimensional corners.

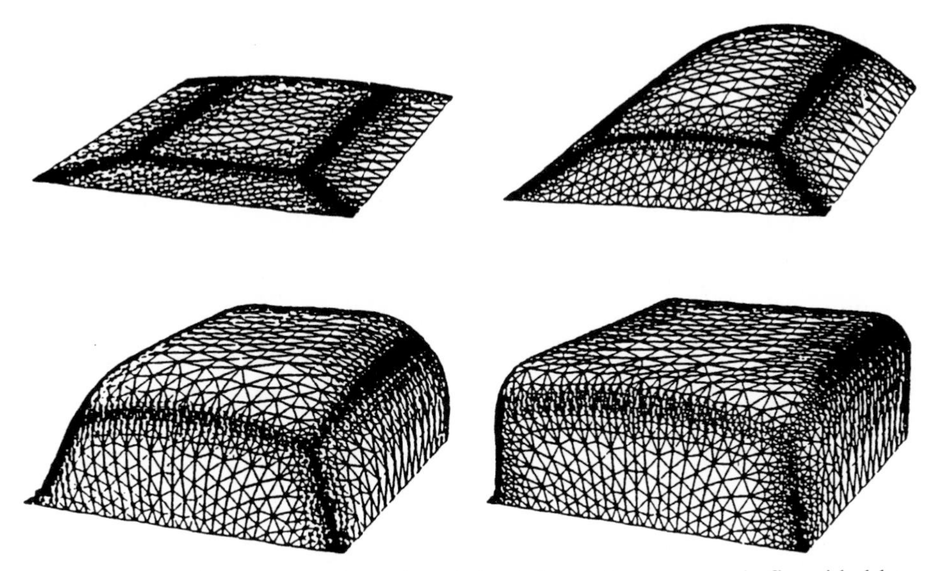

Figure 9.2 FEA computer-time plot of sheet formation over male five-sided box.

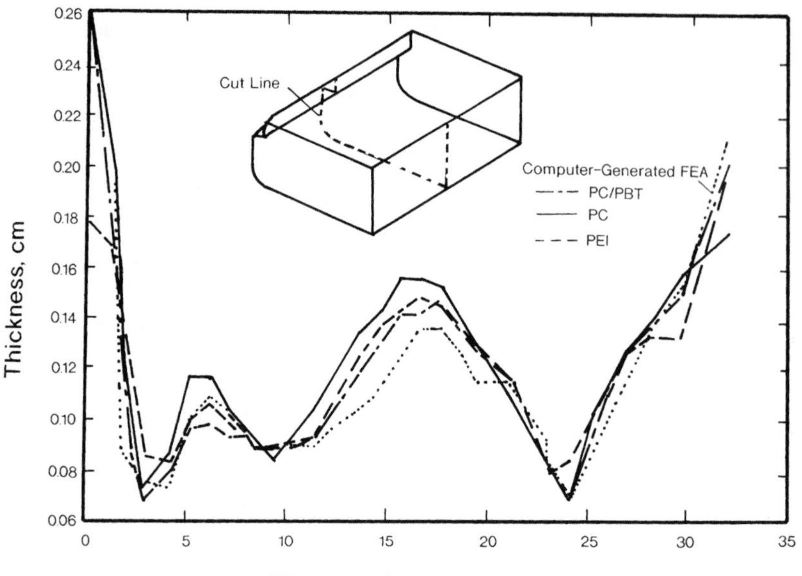

Figure 9.3 Comparison of measured and predicted wall thickness for three thermoformed polymers.

9.4 Some Guides to Successful Part Design

Some aspects of part design such as draft angles and texture characteristics, have been discussed earlier.

- Webbing typically at outside three-dimensional corners on male molds and male portions of female molds, Fig. 9.4. The local sheet surface area is greater than the local area of the mold surface. Webbing is mitigated on male molds with stand-off blocks called web catchers and on male portions of female molds with proper local plug assist design.
- Any draft on a mold surface is better than no draft at all.
- Chamfers should be considered if generous radii cannot be designed in.
- Half to three-quarters of part shrinkage occurs before the part temperature has fallen to the polymer heat distortion value.
- For vacuum forming, the minimum radius in a two-dimensional or three-dimensional corner should be greater than the local sheet thickness.
- A radius four times the sheet thickness is formed easily in most thermoplastics.
- Plastics sheet thins in proportion to the radius of the three-dimensional corner.

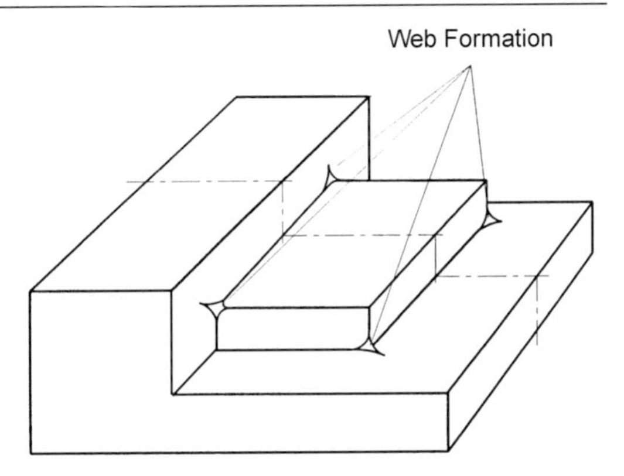

Figure 9.4 Formation of webs on inside three-dimensional corners of rectangular male mold.

- Pressure forming allows corner radii of 0.015 in or 0.4 μm for most polymers.
- Smaller radii are achieved with hotter sheet and faster forming times.
- Chill marks are an indication of rapid local thickness change.
- Shiny spots are an indication of inadequate vacuum locally.
- Some plastics such as PMMA and GPPS are quite notch-sensitive.
- Sharp corners can be brittle or splitty.
- Parts with angles less than 90 degrees may be brittle regardless of the polymer.
- To obtain local part thickness accuracy, the part should be formed with thicker sheet and the region routed to dimension.
- Ribs, corrugations, flutes and multiple cones are typical ways of stiffening thermoformed part walls.
- Large surface area parts are stiffened by adding a slight dome of 15%, concentric ribs, radial ribs, or combinations of these.
- The distances between multiple male ribs should be at least 150% of the height of the ribs. This holds for slots as well.
- Shrinkage of fiber-reinforced parts is less important than dimensional changes due to elastic recovery or "spring-back" once the forming forces are removed.
- Slots on vertical sides of female parts should run parallel to the plane of the sheet.
- Parts should be designed to form around fully extended movable side cores.
- Rolled rims are typically hundreds of times stiffer than flat rims. Small diameter rolled rims on drink cups lead to undesirable wicking.

- Part design inaccuracies are usually caused by
  - inaccuracies in tolling,
  - sheet-to-sheet variations in material properties and dimension,
  - lot-to-lot variations in sheet quality and properties,
  - female versus male tooling,
  - part geometry that compromises formability,
  - the use of less-than-adequate forming technology,
  - poor operator skills, and
  - improper or inadequate maintenance on machines molds.

10 Issues of Quality Control

Success in thermoforming depends on the economic production of products that meet the customer's specifications. Quality control begins with acceptance of quality sheet and ends with parts that meet established acceptable quality limits (AQLs).

10.1 Incoming Sheet Quality

Whether rolled or palletized, sheet is the incoming material for the thermoformer. To assure quality parts, thermoformers must require quality sheet. In addition, they must understand enough about the extrusion process to understand its limitations. These limitations may be machinery-driven, such as maximum or minimum sheet width, or they may be specific for a given polymer.

10.1.1 The Extrusion Process – What a Thermoformer Needs to Know

The basic extrusion process is shown in schematic in Fig. 10.1. Solid polymer, usually a mixture of virgin pellets and regrind, is added to the hopper of the extruder. The barrel of the extruder contains a flighted screw that is turned with an electric motor. The barrel is heated electrically. The plastic is conveyed by the screw down the barrel, where it is heated and melted under pressure. The molten polymer is then squeezed through a shaping die. The gap at the die end is adjusted to provide the desired sheet thickness. The extruded plastic is laid on one roll of a multiroll stack, and serpentined from roll to roll. The rolls are temperature-controlled. The roll speed is matched to the extruder throughput rate to minimize sheet orientation. Heavy-gauge sheet is then cooled prior to guillotine or saw cutting and stacking. Thin-gauge sheet is fed to a take-up winder. Gauge thickness is determined manually for heavy-gauge sheet and with nuclear gauges for thin-gauge sheet.

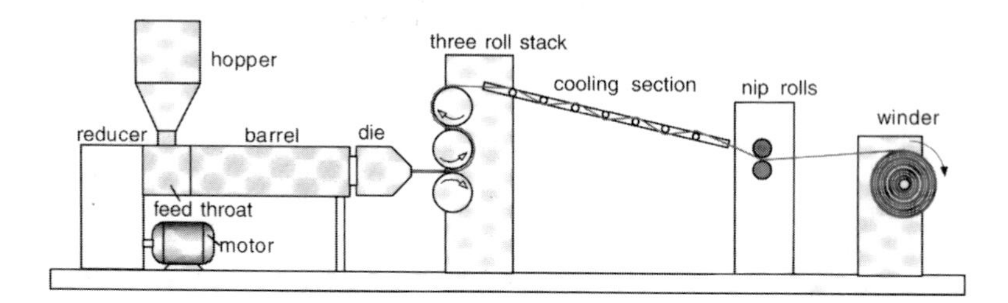

Figure 10.1 Typical sheet or flat film extrusion line.

There are many variations to the process in Fig. 10.1. Twin-screw extruders are used for temperature-sensitive polymers such as PVC. Multiple extruders are used for coextruded, multilayer, and laminated sheet. Tandem extruders are used to produce low-density foams. Hopper dryers are used on many extruders to remove moisture from extruded pellets and regrind chip. Certain polymers such as PET and ABS require substantial drying before processing.

10.1.2 The Purchase Order

The purchase agreement between the extruder and the thermoformer is as important as that between the thermoformer and the customer. The agreement should include the expected amount or number of quality sheets, along with the sheet dimensions of width, length, and thickness. Sheet thickness tolerance is about 1 to 2 mils (0.001 to 0.002 in or 25 to 50 μm) for thin-gauge sheet and about 5% for medium and heavy-gauge sheet. Sheet squareness should be $90^\circ \pm 0.25^\circ$. Out-of-flatness should not exceed 2% of the diagonal dimension of the sheet. In certain cases, an out-of-flatness specification at an elevated temperature may be needed and a suitable test for this requirement agreed on in advance.

If the polymer is prone to moisture pick-up, as are PC, ABS, PET, and PMMA, a drying protocol should be agreed on. Table 10.1 gives some typical drying times for pelletized polymers and regrind chips. If moisture is a serious impediment to successful forming, the thermoformer and extruder need to agree on an appropriate test for moisture in the extruded sheet. When drying sheet that is suspect, the drying times of Table 10.1 are for 40 mil, (0.040 in or 1 mm) of sheet thickness.

Table 10.1 Drying Conditions for Some Thermoformable Polymers

Polymer	Typical Drying Temperature [°F]	Typical Drying Temperature [°C]	Typical Drying Time [hr]
APET	150	65	3–4
CPET	320	160	4
ABS	175	80	2
PBT	320	160	4
PMMA	175	80	3
PC	300	150	4

Sheet appearance is of paramount importance in many forming applications. The level of sheet quality also should be agreed on. Some sheet quality concerns include:

- linear surface marks
- irregular surface marks
- dents
- chatter or ribbon marks
- microscopic webbing
- holes
- pits
- lumps
- specks
- gels
- color uniformity and intensity
- surface appearance, such as level of gloss or uniformity in texture

In certain cases, anti-static agents or anti-blocking agents need to be added at the extruder. The level and uniformity of these agents must also be agreed on in advance of production.

Certain polymers are dramatically affected by thermal and shear processing or moisture. In the presence of even small amounts of moisture, PET, PC, and nylons quickly lose molecular weight when heated to processing temperatures. Material property loss can result in excessive sag, extreme thinning during stretching, or brittle product failure. Excessive heating and shear degrades RPVC. The first indication is yellowing in color. HDPE degrades to lower molecular weight material. This results in excessive thinning during stretching and, to some degree, a loss in UV and chemical resistance. Since HDPE's molecular weight is related to viscosity, a simple melt indexer is used to determine the effects of processing and reprocessing HDPE.

Economically efficient thermoforming depends on the successful reuse of trim. It is imperative that the extruder clearly understands how the thermoformer wants regrind handled. The purchase order should designate the level and source of the regrind stream and the amount of regrind to be used. Regrind is not allowed in certain applications, such as medical packaging. In other applications, such as dunnage, 100% regrind is used. The extruder and thermoformer should be aware that excessive regrind in the extrusion process may cause quality problems in the sheet, such as excessive gels, specks, lumps, poor surface gloss, mottled color, color change, and poor gauge control. Today, many purchase orders state an allowable regrind range, such as "30% (wt) ± 5% (wt) regrind from XYZ company only."

The thermoformer and extruder should agree on the maximum allowable machine and cross-machine direction orientations. They should also agree on a simple test for sheet orientation. Excessive orientation may lead to sheet pull-out from clamp frames or pins. One orientation test begins with 1 in (25 mm) by 10 in (250 mm) strips cut from the sheet. One set of strips are cut with the long sides in the machine direction. Another set are cut with the long sides in the transverse direction. These strips are placed on a talc-powdered metal plate and the assembly placed in a forced air convection oven at the expected forming temperature of the plastic. After several minutes, the assembly is removed, cooled, and the strips measured to determine orientation in both machine and transverse directions. Typically, thin-gauge thermoforming can tolerate greater orientation than heavy-gauge thermoforming. Sheet with no more than 5% orientation is usually satisfactory for most thermoforming applications, but maximum levels depend strongly on the polymer and the application.

Other specifications for given applications may be needed. For example, residual odor is critical in certain food product containers. Odor suppressants are available as additives in many polymers such as PP, but many are thermally sensitive and so care must be taken in heating the polymer. Similar care must be taken with certain organic dyes, fire retardants, and ultraviolet stabilizers.

When fire retardancy is mandated or when an undetected loss in physical properties could result in catastrophic failure, the testing protocol must be formalized to include signed-off data sheets that accompany the shipment. Table 10.2 is a typical checklist of sheet specifications that should be agreed on and appended to the purchase order.

Table 10.2 Sheet Purchasing Specification Check List

Specifications	Certifier/Tester	Comments
Degree of orientation allowed	T	
Sheet sag characteristics	T	Material consideration but extrusion characteristics considered as well
Use of regrind, trim, selvage	T	
Gauge tolerance	T	Sheet-to-sheet accuracy may require extruder input
Width, length, flatness tolerance	T	Extruder input useful
Impact strength [drop ball, dart, Izod]	T	*a priori* decision on who runs test
Moisture level	T/X	Specific drying level required for moisture-sensitive materials
Foreign matter, agglomerations, Type, frequency	T/X	Important for polymers that burn, or discolor, processing aids, fillers, fire retardants, crosslinking
Gel count	T/X	See comments above
Finish surface required		
Texture	T/X	
Smoothness	T/X	
Gloss	T/X	
Pits, dimples, waves, air entrapment, bumps	T	Good products made from quality sheet
Optics	T	
Mechanical properties	T	Translation of polymer properties into sheet responsibility of thermoformer
Pigment distribution	T/X	Type of test must be made *a priori*
Filler condition	T/X	Particle size, drying conditions
Fire retardant condition	T/X	Method of addition, determination of loss of effectiveness
Odor	T/X	
Laminate properties		
Moisture transmission	T/X	Type of test must be made *a priori*
Oxygen permeability	T/X	Type of test must be made *a priori*
Packaging, shipping	T/X	Roll diameter, core size, method of palletizing, protective wrap, moisture protection

T = Thermoformer, X = Extruder, T/X = Both

10.1.3 Incoming Sheet Quality Evaluation

It is apparent from Table 10.2 that substantial effort should be expended to assure that the as-delivered sheet meets the highest possible quality standards. Most thermoformed products do not demand extensive in-coming testing. Visual and dimensional inspections of the sheet are usually done by the machine operator as the sheet is fed into the forming press. Inspection results are usually recorded only in critical applications. Spot checks of polymer quality are done when the product must meet critical design specifications for such properties as fire retardancy or impact strength. Sheet lot retains are recommended for medical and food product containers.

10.2 Production Monitoring

There are three aspects of the thermoforming process that warrant monitoring. Efficient energy input yields sheet that is always at the proper forming temperature. Uniform stretching of the sheet yields predictable, tight tolerance part wall thickness every time. Properly sharpened and guided cutting edges yield fracture surfaces that have a minimum of microcracks and cutter dust.

10.2.1 Measuring Temperature

Many thermoformers exit sheet from the oven based on the time the sheet has spent in the oven. Sheet that exits based on time may not always be heated the same, time after time. Uneven heating can result from aging heaters, burned-out heaters, inefficient reflectors, non-uniform sheet sag, unpredictable air flow through the oven, varying oven air temperature, power surges, and the unpredictable "on-off" sequencing of the heaters. Recently, thermoforming machines have been built with devices that measure infrared temperature directly through the oven wall. In many cases, thermoforming operators simply record the sheet temperature while still exiting sheet on based on time. In heavy-gauge forming with oven residence time control, the sheet is exited based on when the measured temperature is equal to the pre-set required temperature.

When pattern or zonal heating is used, single-point temperature measurements are not always reliable. In certain instances, inexpensive infrared measuring devices are mounted along the path taken by the sheet as it exits the oven. The measured multiple strip temperatures are then combined into a temperature profile of the entire sheet through a computer software program. A more expensive device called a thermal imager takes an infrared video image of the entire surface area of the sheet. Computer software is used to produce a five-or ten-color image that vividly contrasts hot and cold areas.

Temperature measurement of in-production mold surfaces is important if non-uniform cooling is suspected. Infrared thermometers have replaced hand-held surface thermocouples for this measurement. Thermal imagers are also useful. When parts are vacuum or drape formed, thermal images are also used to monitor sheet cooling.

10.2.2 Sheet Formability

Sheet sag is frequently a strong indication that the sheet is nearing its forming temperature. Many variables affect sag, such as sheet weight, sheet span, the type of polymer, heater spacing, extruder machine and cross-direction orientation, filler loading, and even sheet color. As a result, it is difficult to predict formability ahead of time. As discussed in Chapter 2, the forming window is strongly related to the plateau region of the temperature-dependent modulus of the specific polymer. Unfortunately, the equipment needed to

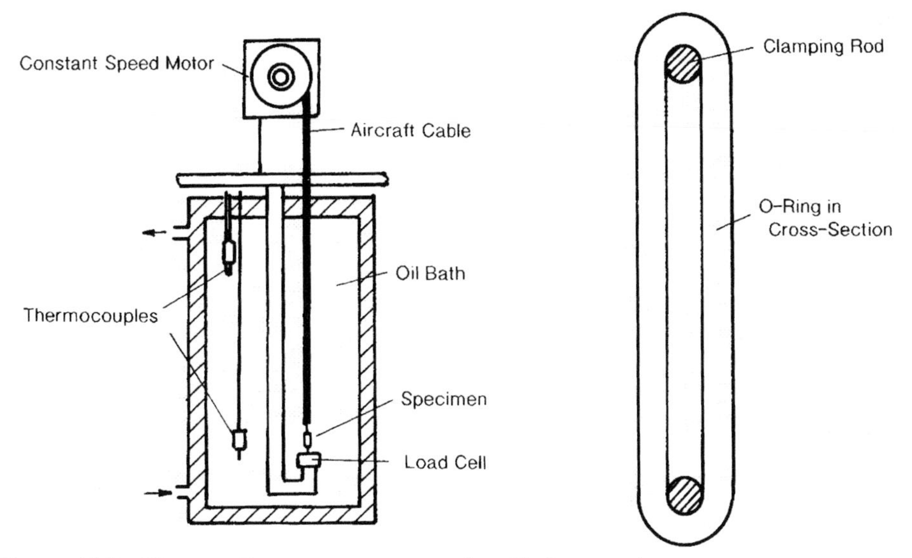

Figure 10.2 Hot tensile test apparatus using O-ring specimen.

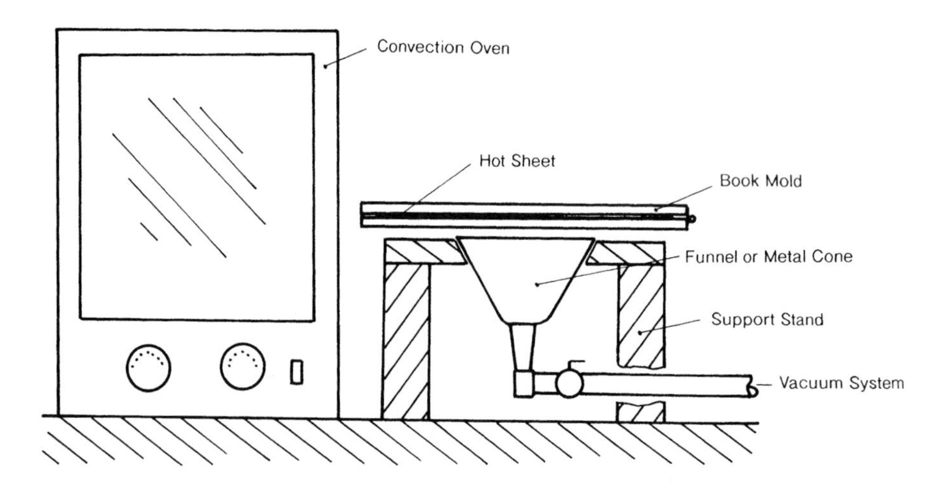

Figure 10.3 Laboratory funnel apparatus for evaluating sheet draw-down capability.

obtain these data is expensive and difficult to operate and maintain. Since the objective is to determine the temperature-dependent stretching characteristics of the polymer, it would appear that a rather simple stretching test would be of use. The hot creep test (Fig. 10.2), yields reasonably useful data. Temperature-dependent maximum draw ratios of forced air convection oven-heated sheet are determined using the 60-degree funnel shown in Fig. 10.3.

Drawdown uniformity is usually determined by cross-hatching a sheet, then heating and forming it. Uniformity is determined by comparing distorted squares of several formed parts. Circle-grids, adapted from sheet metal forming, are also used. Where circles distort to circles, the sheet is stretched uniformly in all directions. Where circles distort to ellipses, the sheet is stretched more in the long direction than in the short direction. The local increase in area is inversely proportional to the local reduction in sheet thickness.

10.2.3 Cutting Surfaces

The easiest way to determine cutting efficiency is to visually inspect the cutting surfaces. Most cutting surfaces can be inspected with inexpensive magnifiers of 8X to 30X. General dulling or rounding of cutting surfaces, as well as nicks and chips, lead to excessive microcracking and cutter dust. These magnifiers are also useful in examining part cut surfaces and cutter dust.

10.2.4 Finished Part Performance

The most obvious question is whether the finished part meets the design specifications. This includes overall dimensions and local sheet thicknesses, where critical. For many applications, thickness gauges and go-no-go fixtures are used.

Coordinate measuring machines (CMMs) are now being used in thermoforming quality control programs. CMMs are basically very accurate, three-dimensional electronic rulers. Beginning from a known reference point, a touch-sensitive pointer is moved across a solid surface and specific significant locations are electronically recorded. Computer software is then used to either log and compare the various locations or to electronically reconstruct the solid surface shape. Early machines were expensive, cumbersome devices that were prone to vary in effectiveness from vibration, air temperature, and humidity.

Newer machines are lightweight and contain compensators for vibration and temperature. The most important use of the CMM is as a quality control device. Critical part dimensions are accurately measured and compared with product specifications. Part-to-part dimensional variations are compared with allowable product variations.

On occasion, the CMM is used to determine dimensional sensitivity to various process and material changes. The CMM is also used to "reverse engineer" an existing part. Part dimensions are used to first electronically reconstruct the part. Mold dimensions are then obtained by adding expected shrinkage values to the measured part dimensions. The CMM is also used to obtain local wall thicknesses. These values are used to determine initial sheet thickness and certain aspects of plug design.

If the product is transparent and is made of PS, PMMA, or to some extent, RPVC, the strain field can be seen as a vivid color field by viewing the part through crossed polarizers. High strain levels are seen as very close color bands. If the product is to be used under conditions of extended high temperatures it is placed in a forced air convection oven at a temperature 20 to 50 °F (10 to 30 °C) above the design temperature to determine potential distortion.

11 Set-Up, Maintenance, and Troubleshooting

The thermoforming process can utilize many different types of machines, all of which require set-up, maintenance, and repair in some fashion. Each person involved with thermoforming should be fully aware of the high temperatures, high pressures, great mechanical forces, pinch-points, high voltage, combustible plastics, and electrically or electronically timed start-stop traveling elements that can maim or kill. A protocol for the set-up, operation, and up-keep of each machine is usually provided by the machinery builder, along with recommended safety practices. These should be carefully reviewed by everyone working on these machines and should be filed in an accessible place for easy reference.

Set-up and maintenance are done when production runs are not occurring. Troubleshooting is frequently done during production. Both are reviewed here.

11.1 Safety

Training is the most important method of promoting plant safety. Accidents often happen because the plant safety program is flawed in some way, workers are not adequately trained or retrained, or the workers ignore safety protocols. Poor housekeeping, exhibited as dangling or frayed extension cords, leaking oil or water, unswept trim dust, blocked aisles and exits, open knives, unattended bandsaws, and even loud radios, is the cause of many accidents. The plant should provide and require workers to wear heat resistant gloves, safety goggles or shields, ear plugs, and any other protective gear recommended by OSHA or the machinery manufacturer.

One common accident occurs when a machine is not completely disabled before workers enter the machine zone. Burns occur if the machine is touched while heaters are still hot, even though the power is off.

Thermoforming and ancillary machines must meet certain safety standards, including those of OSHA. Cages with microswitches, pressure plates, and light curtains are designed to deactivate a functional machine. Prominently displayed cautionary signs in Alert! orange are placed at potential pinch-points, reach-in points, and all electrical boxes.

One important safety program emphasizes "Lock it and Pocket!" Under this program, anyone accessing a machine locks out the machine controls and puts the key in his or her pocket. This prevents anyone else from accidentally turning on the machine. Even then, accidents can happen if heaters are still hot or pneumatic or hydraulic cylinders are still under pressure.

It should never be assumed that everyone knows how to set up, operate, shut down, or maintain every piece of equipment. In many plants, employees are cross-trained, i.e., they are required to work at every major station in the plant. While this is an efficient use of labor, it is imperative that each employee also be fully trained on the safety aspects of each machine at each station. Every employee should be trained to monitor his or her own safety and not rely on advice from coworkers or supervisors. One successful safety program emphasizes "Safety Begins With Me!"

Temporary employees should never be trained "on-the-job." Before any temporary employee is hired, a rigorous indoctrination program designed specifically for the casual employee should be in place and tested by permanent employees.

11.2 Thermoforming Machine Set-Up

There are two types of machine set-up. When a new machine is ordered, the machinery builder should offer an extensive training program on the new equipment. It is imperative that at least three people: an operator skilled on similar equipment, a maintenance person, and the technical person who specified the primary machine elements, take the training program. In many cases, a technical service representative from the machinery builder may instruct plant employees on-site on the machine's operation after installation. The maintenance person should work with the machinery technical service representative throughout the installation to determine any potential problems concerning the new machine. Once power has been delivered to the machine, the operator should then join the training session. The set-up person should also be available at this point.

A mold that has run well in the past should then be installed in the press. Platen level and alignment should be checked. All stops should be set, coolant provided to the mold, and the mold dry-cycled to ensure uniform platen travel and closure. The sheet delivery system should then be activated and the machine function tested without heat. The delivery and press systems should be deactivated and the top oven heaters turned on. When the heaters cycle is under control, individual heater temperatures should be measured and compared with PLC-indicated temperatures. The top heaters should then be shut off, the lower heaters turned on, and the measurements repeated. A check of the vacuum system should then be made.

Next, the mold cavity should be covered with a rigid plate containing a vacuum gauge (Fig. 11.1), and then, the rotary vacuum valve opened. The vacuum system should recover to its previous vacuum value at a rate faster than the expected cooling rate of the formed part. All the safeguards, such as light curtains, safety gates, fire extinguishers, and warning alarms and lights, should then be set and checked.

The former should now be ready to operate. It is recommended that the first parts be vacuum-formed. After the machine is deemed to be operating satisfactorily, pressure box, plug assist, peripheral clamping or gridding, and pre-blow functions included on the mold should then be tested. For thin-gauge forming, in-place or in-press trimming should be actuated only after the forming elements are functioning satisfactorily.

Occasionally, used machines are purchased. It is imperative that the original machinery builder be contacted before the machine is operated. The original machinery builder may supply operating guides, recommended or required safety upgrades, and technical service start-up and training. If the used machine is rebuilt before delivery, the machinery rebuilder may be able to supply missing documents and training guides, as well as offer technical service. If the machine is still operating at the time of purchase, the purchaser should contract training time on it before it is dismantled for shipment. Under any conditions, everyone starting,

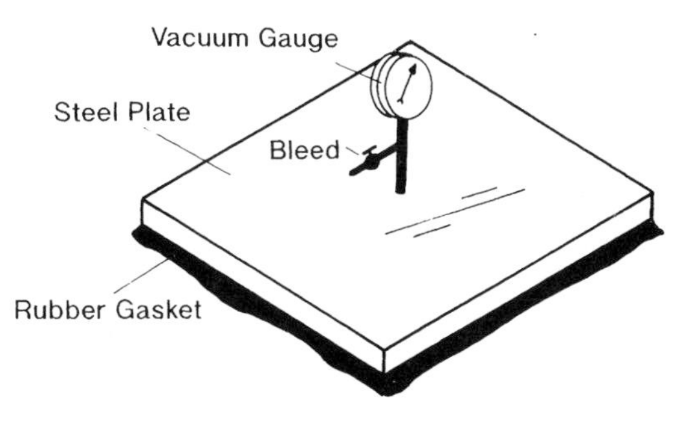

Figure 11.1 Device for checking vacuum capacity for given mold cavity design.

operating, or maintaining a used machine should be fully trained in its function prior to operating it in its new location.

Set-up on an existing machine usually entails mold replacement, along with ancillary features such as pressure box, plug assist, in-place or in-press trimming fixtures, and stacking elements. If mold change-out is a major down-time factor, efforts to speed set-up are recommended. The extra expense in providing a modular mold assembly that includes plug assist, peripheral or grid clamps, vacuum box, in-place trim dies, and even pressure box and sheet clamp frame, may be repaid in rapid mold change-out time. In an established plant, there is some reluctance to start using modular mold assemblies because older molds are not similarly modular. However, the expense of retrofitting frequently used older molds may frequently be justified in up-time improvement. Machine parameters should always be electronically stored. Most PLC-based forming machines have data storage capacity.

11.3 Mold Set-Up

As with thermoforming machines, there are two types of mold set-up. There needs to be substantial dialogue between the mold builder and the buyer from the time a new mold is ordered until it is ready for acceptance. Assuming that this has been done, the buyer now needs to make certain the mold is ready to run prior to accepting delivery. The mold surface should be visually inspected for texture and finish. Coolant lines should be chip-free and properly laid out, then pressurized with water to make certain there are no pinholes, perforations or errant vent holes. All mechanical actions, including slides, plug, grids, ejectors, locking bayonets, and guide-pins should be checked to ensure smooth functioning. All vacuum holes should be open. If the mold includes in-place trimming, trim action should work smoothly and steel rule die adjustment should be easy. Pick and lift points should be clearly marked. Elements such as logos and date-stamps should be keyed for easy replacement.

The most frustrating part of setting up an existing mold is not having all the necessary pieces at the time of installation. The use of modular mold assemblies helps mitigate this problem, as do accurately recorded press set-up details from the previous installation. Previously recorded data are particularly important when set-up requires critical positioning of plug assists or gapping of in-place trim dies.

11.4 Trim Set-Up

Thin-gauge horizontal or canopy trim presses are notoriously difficult to set up, primarily because of the mass of the punch-and-die assemblies. The time to "make-ready" is lessened by using modular punch-and-die assemblies, but a substantial amount of time is still spent in the manual adjustments necessary to bring the press in. This set-up time is doubled when tandem trim presses are used.

In-machine trim stations that are not modular are equally difficult to set up. Care must be taken to ensure complete die-to-anvil contact. Tissue or kraft paper is frequently used as a measure of complete contact. If the die cuts into a gap or slot rather than on an anvil, there cannot be interference or the die will quickly wear unevenly. Set-up must allow for thermal expansion of the die if it is heated during production. Mechanical shimming is frequently used, although more expensive trim dies incorporate micrometer adjustments.

Today, the set-up of a heavy-gauge trimmer means tool-path programming of the multi-axis machine. This effort can be laborious and is frequently not user-friendly. Once the program is verified, it can be recalled from memory at any time. It must be remembered that the multi-axis machine always assumes that each part is fixtured exactly at the same X-Y-Z coordinates as the ideal one stored in the computer. Subtle part dimensional changes, minor changes in the position of the fixture, or variation in the way the part fits on the fixture affects the accuracy of the trim line.

At least one multi-axis trim machine builder recommends that a "16 point check" be made prior to resetting any fixture. The tip of a specific cutter is positioned at a fixed X-Y-Z coordinate. The machine then moves in sequence to each major point on the horizontal compass, each major point on the vertical compass, and each major point on circles at 45° to the horizontal and vertical circles. The set-up person observes how much the tip of the cutter moves as the machine moves through these 16 stations. Because this point is internal to the machine, it can always be used as the reference point when resetting a fixture.

The trim fixture and a near-ideal trimmed part should always be stored with the mold. When the fixture is reinstalled, the trimmed part should be refitted over it and the trim program initiated to ensure that the trim path is correct. Since the fixture must hold the part rigidly throughout the trimming procedure, care must be taken to ensure that all the vacuum holes are functional and air does not leak in as the trim is machined away. Parts that are slightly oversized may not be held tightly against the fixtures. Air leak may pose potential loss of dimension during trimming. Slightly undersized parts are sometimes hammered onto fixtures. This imparts additional stresses that may interfere with the performance of the part.

11.5 Maintenance

There are two general types of maintenance performed in modern plants. The first is emergency or crisis maintenance, where unforeseen and unpredicted elements fail. The second is preventative maintenance. In preventative maintenance, a routine of inspection, repair, and replacement is established, for the purpose of preventing the need for the first type of maintenance, i.e., as a result of catastrophic machine failure.

Many plants schedule periodic line shut-downs for preventative maintenance. For instance, once a year, during an expected slow time, a machine may be completely shut down, inspected, and repaired. Thermoforming machine maintenance should include:

- complete efficiency testing of all heaters
- press frame inspection for rust and mechanical breaks
- inspection of sheet transfer elements for excessive wear
- testing and repair of pin-chains or pneumatic clamp cylinders
- inspection of all electrical contacts for gap erosion and sticking
- complete inspection of all hoses and connections on the press, including all overhead and pit connections
- examination and replacement of hydraulic fluids
- replacement of pressure box seals
- disassembly of the vacuum pump for inspection of rotors and replacement of worn parts

Platens should be leveled and checked for parallelism. All press motions should be carefully monitored for smooth action and all metal-to-metal contacts should be examined for wear.

Likewise, trim presses should be examined for structural integrity and excessive wear. Above all, no machine should be put back into service until all oil and water leaks are repaired and the machine cleaned of all crud, including grease, trimdust, and plant dirt. Rust should be scraped and the metal primed and painted. At least 50 hours should be budgeted for this major production line overhaul.

Monthly programs are recommended in addition to annual programs. Usually this effort takes about 5 to 10 man-hours and focuses on replacing reflectors; detecting and replacing burned-out heaters; examining steel rule dies and punch-and-die sets for nicks, dulling and wear; changing vacuum pump oil; and draining water from the surge tank.

For medium to long production runs, weekly preventative maintenance should focus on general lubrication and data down-loading. Heater operation should be monitored if very sensitive heater elements are used. If the polymer is particularly difficult to trim, steel rule dies or router bits should be examined for dulling and wear. This is usually a good time to wipe excess grease, trim dust, plant dirt, and other crud from the machine, and to clean the floor and machine areas of detritus. A weekly program should take a couple of hours. For short production runs, preventative maintenance should be budgeted into mold change-out time. Daily maintenance, particularly in the trimming area, is sometimes needed for difficult-to-run plastics.

Mold maintenance should be done whenever the mold is changed out of the press. A complete inspection should include:

- microscopic examination of the critical portions of the mold surface to determine porosity, microcracks, loss of texture sharpness
- examination of moat and dam areas for wearing
- examination of the nature of deposits and buildup on the mold surface, followed by careful cleaning
- examination of water lines for deposits and buildup
- examination of slide cams, hinges, guide pins, plugs, grids, and ejector areas for sticking or excessive wear

Air should be blown back through vent holes to ensure that they are clear of detritus. All vent holes should be microscopically examined to determine if they are still round. If not, holes should be redrilled.

Plugs should be inspected for compression damage. After repeatedly being pressed into a heated sheet, the plug may not retain its original shape. If the plug is flattened, it should be replaced. If the plugs are coated, the coating should be carefully examined for wear. If the coating is worn away, it should be replaced, or a new plug obtained. The mold, an acceptable part, and all the ancillary pieces that are used to make the part, should then be stored in a dry area. If wooden molds or plugs are used, they should be stored in an air-conditioned area to minimize checking or cracking.

Crisis maintenance obviously takes priority over any other program, including set-up, mold change-out, and preventative maintenance. The keys to successful crisis management are rapid identification of the cause of failure and the rapid replacement of the failed part. Although keeping inventories of certain critical parts on hand is prudent, it is not necessary to have a replacement part for each part on the machines. Instead, extensive lists of parts and sources for them should be easily accessible to those in charge of getting the operation functional. Many parts suppliers have same-day delivery service and even may be able to recommend local sources.

In certain instances, patchwork repair may allow the operation to limp along until repair parts are received and installed. Under all conditions, a careful examination of the failed part must be made to determine the cause of failure. If the part is too large to keep, photographs of the failed areas must be taken. If a specific type of failure is repeated, the supplier of the part and/or the machine should be alerted. Catastrophic failure may be relatively easy to fix, but occasionally may take days. This approach to crisis management is also applicable to mold and trimmer crises.

11.6 Troubleshooting

It has been said that there is never enough time to do a project right the first time, but there is always enough time to do it over. Troubleshooting is the term used to describe the process of solving problems. There are many potential sources for problems in thermoforming, beginning with conceptual design and continuing through polymer selection, extrusion, sheet handling and storage, mold design, heating, forming, trimming, and ending with inspection and shipping.

The standard approach to problem solving begins with a clear definition of the problem. For example, linear lines in one direction are found in a formed part. Inspection determines that these lines are also in the sheet. Next, a probable cause is identified. In the example, die lines in extrusion are caused by build-up of detritus in the extrusion die. Then a course or courses of action are proposed. In this case, the die needs to be thoroughly cleaned. If the problem reoccurs, the cause of the plate-out needs to be determined.

Many thermoforming troubleshooting guides are available from machinery builders, extruders, polymer suppliers, consultants, and source books. Each identifies the specific problem, as listed in Table 11.1. Then various probable causes and suggestions for eliminating or removing the causes are given. In certain cases, the possible causes are usually categorized as resulting from problems in production, tooling, machinery, polymer materials, or design. As an example, the formed part is found to have local color change in corners. A production cause may be cold sheet or cold plug. The production action would be to raise sheet temperature or heat the

Table 11.1 Some Thermoformed Part Process/Product Problems

Blisters or bubbles	Poor forming and bad detail
Distorted product surface	Color changes
White marks	Webbing, bridging or wrinkling
Mold side bumps	Chill marks
Surface marks	Locally shiny surface
Excessive post-forming distortion	Excessive post-forming shrinkage
Warped or twisted part	Thin corners
Locally thin surfaces	Thin sides
Poor wall thickness distribution	Post-forming cracking
Shrink marks	Excessive sag
Pre-blow/vacuum bubble variation	Material sticking to plug assist
Material tearing during forming	Material sticking in mold
Edge splits during use	Parts crack in corners during use

plug. A molding or tooling cause may be insufficient draft or very small corner radii. The tooling action would be to increase draft, use greater radii, or replace the radii with chamfers. A machine cause may be inefficient heaters, and the action to correct the problem is to replace reflectors or heaters. The polymer may have poor extensional properties, allowing it to thin too much at high levels of stretch. The solution would be to use a polymer with greater resistance to elongation.

In certain instances, the recommended actions may be contradictory. For example, webbing is an unaccepted fold of polymer, typically in three-dimensional corners of male portions of molds. In certain instances, the mold may be too hot, but in others, it may be too cold. In this case, the problem solver should look to other causes and actions before determining if tooling temperature is an issue.

There are some general rules to follow when troubleshooting. First, return to basics. That is, there are too many process variables involved if a pressure box, peripheral or grid clamping, prestretching, pattern or zonal heating, and in-place trimming are all employed. The basic thermoforming process is vacuum or drape forming, and it is strongly recommended that all auxiliary processes be disabled and the heating pattern be returned to uniform sheet temperature before beginning the troubleshooting process. Then, although auxiliary process effects are not additive, each one should be added back to the process, one at a time, to determine whether it may be the source of the problem.

Second, it has been estimated that fully three-quarters of all thermoforming processing problems are material problems. Possible causes range from unacceptable sheet orientation to improper choice of polymer for the application to the inappropriate use of regrind. Since this cause is so prevalent, the sections on incoming sheet quality control in Chapter 10 should be frequently reviewed.

And finally, troubleshooting is deductive reasoning. It requires instinct, observation, inference, imagination, and systemized common sense. There is a cause for every problem. It is important to consider everything that might be the cause, then carefully eliminate items, one at a time. Recall the Sherlock Holmes adages that the little things are the most important and that when the impossible has been eliminated, that which remains, however improbable, must be the cause. Also keep in mind that the most mundane problem may be the most difficult to solve, because it offers very little from which deductions can be made.

Glossary of Thermoforming Terms

A

Amorphous polymer	Polymer that has no melting temperature
Absorption	The amount of radiant energy absorbed by plastic
Angel hair	Fine fibers caused by improper trimming technique
Areal draw ratio	The ratio of the surface area of the formed part to that of the sheet used to form the part

B

Billow	Pre-stretching sheet by inflation with air pressure
Blend	Physical melt-mixing of two or more polymers

C

CMM	See Coordinate measuring machine
Cavity isolator	See Grid
Chamfer	Bevel
Chill mark	A mark on the formed part that is often attributed to contact with cold mold or plug
Coining	Localized pressing of heated sheet between two portions of mold
Conduction	Energy transfer by direct solid-to-solid contact
Contact heating	Heating of sheet by conduction
Convection	Energy transfer by moving or flowing fluids
Coordinate measuring machine	Accurate three-dimensional electronic ruler used in quality control
Copolymer	Polymer with two sets of monomers, such as HIPS
Cross-machine direction	At right angles to the extrusion direction; also known as transverse direction
Crystalline polymer	Polymer that exhibits melting temperature

Cut-sheet	Usually, heavy-gauge sheet, fed one at a time to a rotary or shuttle thermoformer
Cutter	Mechanical bit with tip designed to cut specific types of plastics

D

Dam	Continuous ridge around mold cavity periphery
Density	Weight per unit volume
Diaphragm forming	Stretching heated sheet using an elastic membrane
Differential pressure	The difference in the pressure on each side of a sheet
Dimensional tolerance	Part-to-part variation in local dimension
Draft	Mold angle from vertical
Draw box	Empty box into which heated sheet is pre-stretched prior to forming
Draw ratio	Measure of the extent of sheet stretching; a measure of the area of the sheet after being formed to that before forming

E

Elastic liquid	A molten polymer that has both fluid and solid characteristics; sometimes called a viscoelastic polymer
Elastic modulus	See Modulus
Enthalpy	A thermodynamic measure of the intrinsic heat content of a polymer
Equilibration	Allowing a sheet to approach uniform temperature throughout, after the heating source is removed
Endothermic foaming agent	A chemical that requires heat to decompose to produce gas for agent foaming; sodium bicarbonate is a common endothermic foaming agent
Energy dome	When energy input to sheet is uniform, temperature at the center of the sheet is higher than that at edges and corners
Exothermic foaming agent	A chemical that gives off heat when decomposing to produce gas for foaming; azodicarbonamide (AZ) is a common exothermic foaming agent

Extrusion	The process of producing sheet
Extrusion/forming line	A process where the extruder output feeds directly into the thermoformer

F

FEA	Finite Element Analysis, a mathematical method for determining stress distribution when an object is mechanically deformed
FEM	Finite Element Method. See FEA
FFS	Form, Fill and Seal; in-line thin-gauge process used in food and medical device packaging
Free surface	The sheet surface that is not in contact with the mold surface
Foaming agent	Additive that produces gas during extrusion to produce foamed sheet
Female mold	A cavity into which the heated sheet is stretched; also known as negative mold

G

Gel	Hard, resinous particle in plastic sheet
Glass transition temperature	The temperature range above which a brittle or tough polymer becomes rubbery
Grid	A mechanical frame that presses hot sheet against a multi-cavity mold; also known as cavity isolator or hold-down grid

H

H:D	Height-to-diameter ratio; a measure of draw ratio
Heat capacity	A measure of the amount of energy required to raise a polymer's temperature a specific amount
Heat transfer coefficient	A measure of the effectiveness of energy transport between a flowing fluid and a solid surface
Heavy-gauge	Commonly, a sheet with thickness greater than 120 mils (0.120 in or 3 mm)
Homopolymer	A polymer made from a single monomer, such as PS

I

Infrared radiation	Electromagnetic energy transmission at wavelengths above visible wavelengths
In-line trimming	In thin-gauge, roll-fed forming, trimming that takes place in a separate machine after the thermoforming machine
In-machine trimming	Trimming that takes place while the formed sheet is still within the thermoforming machine
In-place trimming	Trimming that takes place while the formed sheet is still on the mold surface

L

Linear draw ratio	The ratio of the length of a line scribed on a formed part to the length of a line scribed on the unformed sheet
Linear expansion	Increase in polymer dimension on heating
Loft	Expansion of fiber-reinforced sheet during heating

M

MDF	Medium-density fiberboard
Machine direction	In the extrusion direction
Male mold	A mold over which the heated sheet is stretched; also called positive mold
Mark-off	A mark on the formed part that is attributed to contact with plug
Matched die forming	The process of forming sheet between two mold halves, commonly used in foam forming
Melt temperature	The temperature range above which a crystalline polymer changes from a rubbery solid to an elastic liquid
Moat	A continuous groove around mold cavity periphery
Modulus	The initial slope of the stress-strain curve for a given polymer
Molding area diagram	Pressure and temperature restrictions overlaid on stress-strain curves for given polymer

N

Neat	Polymers that contain no additives, fillers, or reinforcing fibers
Negative mold	See Female mold

O

Orientation	The amount of residual or frozen-in strain or stretch in a plastic sheet, usually in a given direction

P

Pattern heating	The practice of selectively shielding portions of heaters to achieve a specific energy input pattern on a heating sheet; also called zoned or zonal heating
Peripheral seal	Region around periphery of twin-sheet part
Pin-chain	Chain that has spikes or pins regularly spaced along its length to impale or hold thin-gauge sheet
Plastic	A mixture of polymers and various additives
Plug	A mechanical device used to aid or assist sheet stretching prior to total contact with the mold
Positive mold	See Male mold
Pressure forming	Commonly, differential pressure across the sheet in excess of 30 lb/in^2 (0.2 MPa or 2 atm)
Punch-and-die	A trimming assembly for thin-gauge forming
Pusher	See Plug

R

Radiation	Electromagnetic energy transfer or interchange between hot and cold surfaces
Reflector	A shaped surface that refocuses radiant energy from a round heater
Replication	Faithful imaging of the mold surface by the hot formed sheet
Residual stress	Frozen-in orientation in sheet
Roll-fed	Thin-gauge sheet, fed continuously into the thermoformer
Rotary press	Heavy-gauge thermoforming machine in which sheet is conveyed from station to station in carrousel fashion

Router	A device for trimming heavy-gauge parts

S

Sag	Distortion of a sheet, under its own weight, while heating
Set temperature	The temperature below which a part can be removed from the mold without appreciable distortion
Shrinkage	Temperature-dependent volumetric change in the polymer
Shuttle press	Heavy-gauge thermoforming machine in which sheet or oven move in a to-and-fro fashion
Skeleton	Trim in thin-gauge forming
Soak time	See Equilibration
Specific heat	See Heat capacity
Specific volume	Volume per unit weight; reciprocal of Density
Spring-back	Elastic recovery of sheet after forming forces are removed
Strain	Stretch; the polymer's response to applied stress
Stiffness	The product of polymer modulus and part geometry
Steel rule die	Sharpened metal band used in compression cutting
Stress	Applied load or force on a sheet, per projected area
Surge tank	The tank between the vacuum pump and the mold assembly to allow near-uniform differential pressure to be applied to the sheet during forming
Syntactic foam	A mixture of sintered inorganic foam spheres and plastic foam matrix, commonly epoxy or polyurethane; used for plugs, fixtures and prototype molds

T

Tab	An uncut portion of a trimmed part, used to retain the part in its web in thin-gauge thermoforming
Terpolymer	A polymer with three sets of monomers, such as ABS

Thermal conductivity	A measure of the time-independent energy transmission through a material
Thermal diffusivity	A material property measure of the rate of energy transmission
Thermoforming window	Temperature range over which sheet is sufficiently soft to allow stretching
Thermoplastics	Two-dimensional organic molecules that can be reprocessed
Thermosets	Three-dimensional organic molecules that cannot be reprocessed
Thick-gauge	See Heavy-gauge
Thin-gauge	Commonly, sheet thickness less than 60 mils (0.060 in or 1.5 mm)
Trapped sheet forming	Conduction heating of sheet
Trim	That portion of the formed sheet that is not part of the final product

V

Vacuum hole	See Vent hole
Vent hole	Small diameter hole drilled through mold surface to allow exhaustion of cavity air; also called vacuum hole
View factor	A measure of the fraction of radiant interchange that occurs between primary heaters and energy absorbers
Virgin polymer	Unprocessed polymer

W

Watt density	Heater output, in W/in^2 or kW/m^2
Wavelength	A measure of the nature of incident electromagnetic radiation
Web	In forming, a fold of plastic that cannot be stretched flat against a mold surface
Web	Trim in thin-gauge thermoforming

Z

Zoned heating	See Pattern heating

Recommendations for Further Reading

Books

Florian, J., *Practical Thermoforming: Principles and Applications*, 2nd Ed., (1996), Marcel Dekker, Inc., New York.

Gruenwald, G., *Thermoforming: A Plastics Processing Guide*, 2nd Ed., (1998), Technomic Publishing Co., Inc., Lancaster PA.

Schwartzmann, P.,/Adolf Illig, *Thermoformen in der Praxis*, (1997), Carl Hanser Verlag, Munich.

Throne, J. L., *Technology of Thermoforming*, (1996), Hanser/Gardner, Munich and Cincinnati.

Throne, J. L., *Thermoforming*, (1986), Carl Hanser Verlag, Munich.

Other Sources

Thermoforming Conference, Society of Plastics Engineers Thermoforming Division, third week of September each year.

Thermoforming Quarterly, published quarterly by Society of Plastics Engineers Thermoforming Division.

Index

N

S

U

V

W

X-Y-Z